International Space Station Benefits for Humanity
2nd Edition

This book was developed collaboratively by the members of the International Space Station Program Science Forum, which includes the National Aeronautics and Space Administration (NASA), Canadian Space Agency (CSA), European Space Agency (ESA), Japan Aerospace Exploration Agency (JAXA), Russian Federal Space Agency (Roscosmos), and the Italian Space Agency (ASI).

NP-2015-01-001-JSC

Acknowledgements

A Product of the ISS Program Science Forum

National Aeronautics and Space Administration:
Julie Robinson, Pete Hasbrook, Amelia Rai, Tara Ruttley, Camille Alleyne, Cynthia Evans, William Stefanov, Michael Read, Kirt Costello, David Hornyak, Tracy Thumm, Susan Anderson, Joshua Byerly, Joshua Buck

Canadian Space Agency:
Nicole Buckley, Luchino Cohen, Ruth Ann Chicoine, Christine Giguère

European Space Agency:
Martin Zell, Eric Istasse, Jason Hatton, Jennifer Ngo-Anh, Nigel Savage, Jon Weems

Japan Aerospace Exploration Agency:
Shigeki Kamigaichi, Kazuyuki Tasaki, Sayaka Umemura, Koki Oikawa, Hideyuki Watanabe, Nobuyoshi Fujimoto, Masato Koyama, Yayoi Miyagawa, Tatsuya Aiba, Shiho Ogawa, Toshitami Ikeda

Russian Federal Space Agency:
Georgy Karabadzhak, Elena Lavrenko, Igor Sorokin, Natalya Zhukova, Nataliya Biryukova, Mark Belakovskiy, Anna Kussmaul

Italian Space Agency:
Salvatore Pignataro, Jean Sabbagh, Germana Galoforo

Executive Editor:
Julie Robinson, NASA

Managing Editor:
Amelia Rai, NASA

Section Editors:
Camille Alleyne, Kirt Costello, David Hornyak, Michael Read, Tara Ruttley, William Stefanov; NASA

Technical Editor:
Neesha Hosein, DB Consulting Group, Inc.

Graphic Designer:
Cynthia Bush, DB Consulting Group, Inc.

Human Health

Earth Observation and Disaster Response

Innovative Technology

Global Education

Economic Development of Space

About the Cover Art

Several themes are integrated into the cover art for this 2nd *International Space Station Benefits for Humanity*. The most central theme is that of "bringing light to the darkness," as indicated by the predominantly nighttime view of Earth, with both the station crew and the populace over which it soars on the verge of experiencing a new dawn. The Earth view chosen includes some of the most underdeveloped regions on the planet; home to people whose lives stand to gain the greatest enrichment from the groundbreaking research being conducted high overhead.

The double-headed arrow, which can be interpreted to be a stylized representation of the space station's orbital flight path, is also the visual centerpiece for two of the remaining themes. The double arrow points both behind the station—back toward Earth, and ahead of the station—forward past the Earth, to the moon, Mars, and to far more distant destinations. This highlights the dual charter of station research to improve life on Earth and to lay the technological groundwork for human expansion beyond low-Earth orbit.

The arrow's color shift from silver to gold represents the third theme: "Invest silver, get back gold." This theme pays homage to the manifold benefits of knowledge realized through international investment in the space station. It also celebrates "gold" of another kind, the enrichment of life on Earth that is the subject of this volume. These benefits for humanity are represented by five gold medallions, each signifying a major area of emphasis: Human Health, Earth Observation and Disaster Response, Innovative Technology, Global Education, and Economic Development of Space.

The final theme is also the most subtle. Although this is a publication coordinated by NASA, the NASA logo is absent from the front cover, in deference to the fact that the benefits chronicled herein are a result, not of any single nation's efforts, nor those of one nation above others, but of unprecedented international partnership and cooperation. This is a benefit for humanity in and of itself, one which promises incalculable rewards not only in such tangibles as heightened international commerce, but perhaps most importantly, in greater cross-cultural understanding and tolerance, the very foundation for humanity's future.

Michael C. Jansen
December 2014

Book Highlights

Robotic arms lend a healing touch
The world's first robotic technology capable of performing surgery inside magnetic resonance machines makes difficult surgeries easier or impossible surgeries possible.
Page 3

Improved eye surgery with space hardware
An eye-tracking device allows the tracking of eye position without interfering with a surgeon's work during corrective laser eye surgery.
Page 6

Bringing space station ultrasound to the ends of the Earth
Small ultrasound units, tele-medicine and remote guidance techniques make medical care more accessible in remote regions.
Page 8

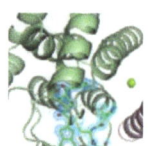
High-quality protein crystal growth experiment aboard Kibo
Protein crystal growth experiments contribute to the development of medical treatments. JAXA is making positive advancements in research on obstinate diseases through experiments in space.
Page 23

Earth remote sensing from the space station
ISS contributes to humanity by collecting data on global climate, environmental change, and natural hazards using its unique complement of crew-operated and automated Earth observation payloads.
Page 51

Advanced ISS technology supports water purification efforts worldwide
At-risk areas can gain access to advanced water filtration and purification systems affording them clean drinking water.
Page 65

Tomatosphere™: Sowing the seeds of discovery through student science
This award-winning educational project with an estimated 3 million students participating is helping researchers answer questions about growing food in space while teaching students about science, agriculture and nutrition.
Page 94

Calling cosmonauts from home
Currently aboard the Russian segment of the station are four space investigations that have educational components to inspire future generations of scientists, technologists, engineers and mathematicians.
Page 104

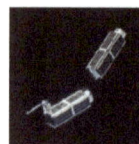
Commercialization of low-Earth orbit (LEO)
Forward-thinking, agile companies like NanoRacks and UrtheCast believe routine utilization of the unique environment of outer space has come of age, and that at long last ISS is open for business.
Page 112

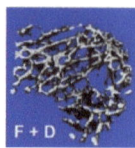
Space mice teach us about muscle and bone loss
Biotech and pharmaceutical companies like Amgen use spaceflight to study their drugs and do preclinical work important for FDA approval.
Page 129

Table of Contents

Acknowledgments ii
About the Cover Art iii
Book Highlights iv
Executive Summary ix
Introduction x

Human Health   1

Health Technology 3
Robotic arms lend a healing touch 3
Robots from space lead to one-stop breast cancer diagnosis treatment 5
Improved eye surgery with space hardware 6
Sensor technologies for high-pressure jobs and operations 7
Bringing space station ultrasound to the ends of the Earth 8
Are you asthmatic? Your new helper comes from space 10
Cold plasmas assist in wound healing 10

Preventing Bone Loss 13
Preventing bone loss in spaceflight with prophylactic use of bisphosphonate:
Health promotion of the elderly by space medicine technologies 13
Improved scanning technologies and insights into osteoporosis 15
Good diet, proper exercise help protect astronauts' bones 15
Add salt? Astronauts' bones say please don't 17

Immune Defenses 19
Early detection of immune changes prevents painful shingles in astronauts and
in Earth-bound patients 19
Station immunology insights for Earth and space 20
Targeted treatments to improve immune response 21

Developing New Therapies 23
High-quality protein crystal growth experiment aboard Kibo 23
Cancer-targeted treatments from space station discoveries 25
Using weightlessness to treat multiple ailments 26

Food and the Environment 29
Microbiology applications from fungal research in space 29
Plant growth on ISS has global impacts on Earth 30
Experiments with higher plants on the Russian Segment of the International Space Station ... 32

Heart Health and Biorhythms 35
   Space cardiology for the benefit of health care 35
   Biological rhythms in space and on Earth 36
   Innovative space-based device promotes restful sleep on Earth 37

Improving Balance and Movement 39
   New technology simulates microgravity and improves balance on Earth 39
   New ways to assess neurovestibular system health in space also benefits those on Earth 40
   Space research leads to non-pharmacological treatment and prevention of vertigo, dizziness
   and equilibrium disturbances 42
   Capturing the secrets of weightless movements for Earth applications 44
   Space technologies in the rehabilitation of movement disorders 45

Earth Observation and Disaster Response   49

Environmental Earth Observations 51
   Earth remote sensing from the space station 51
   Coastal ocean sensing extended mission 53
   Visual and instrumental scientific observation of the ocean from space 54

Disaster Response 57
   Space station camera captures Earthly disaster scenes 57
   Clear high-definition images aid disaster response 59

Innovative Technology   63

Fluids and Clean Water 65
   Advanced ISS technology supports water purification efforts worldwide 65
   Exploring the wonders of fluid motion:
   Improving life on Earth through understanding the nature of Marangoni convection 66
   Space station-inspired mWater app identifies healthy water sources 68
   Space-tested fluid flow concept advances infectious disease diagnoses 69

Materials 71
   Improving semiconductors with nanofibers 71
   InSPACE's big news in the nano world 72

Satellites 75
   Deploying small satellites from ISS 75
   Pinpointing time and location 77
   Space station technology demonstration could boost a new era of satellite-servicing 78

Transportation Technology 81
   Cool flame research aboard space station may lead to a cleaner environment on Earth 81

Robotics 83
   Robonaut's potential shines in multiple space, medical and industrial applications 83

Global Education   87

Inspiring the next generation of students with the International Space Station 89

Inquiry-based Learning 91
Student scientists receive unexpected results from research in space 91
Europe's alliance with space droids 92
NASA has a HUNCH about student success in engineering 93
Tomatosphere™: Sowing the seeds of discovery through student science 94
Students photograph Earth from space via Sally Ride EarthKAM program 96
Try zero G 2: Igniting the passion of the next generation in Asia 97

Inspiration 99
Asian students work with astronauts in space missions 99
Educational benefits of the space experiment "Shadow-beacon" on ISS 100
Students get fit the astronaut way 102
Inspiring youth with a call to the International Space Station 103
Calling cosmonauts from home 104
MAI-75 experiment, main results and prospects for development in education 105

Economic Development of Space   109

Commercial Service Providers 111
Water production in space: Thirsting for a solution 111
Commercialization of low-Earth orbit (LEO) 112
Innovative public-private partnerships for ISS cargo services: Part 1 113
Innovative public-private partnerships for ISS cargo services: Part 2 115
Precision pointing platform for Earth observations from the ISS 116
The Groundbreaker: Earth observation 118
A flock of CubeSats photographs our changing planet 119
Stretch your horizons, Stay Curious™ 120
Mission critical: Flatworm experiment races the clock after splashdown 122
Economic development of space in JAXA 123

Commercial Research 127
Colloids in space: Where consumer products and science intersect 127
Space mice teach us about muscle and bone loss 129
Protein crystals in microgravity 130
Muscle atrophy: Mice on the ISS helping life on Earth 131

Link to Archived Stories and Videos 134

Authors and Principal Investigators by Section 135

Executive Summary

The International Space Station (ISS) is a unique scientific platform that enables researchers from all over the world to put their talents to work on innovative experiments that could not be done anywhere else. Although each space station partner has distinct agency goals for station research, each partner shares a unified goal to extend the resulting knowledge for the betterment of humanity. We may not know yet what will be the most important discovery gained from the space station, but we already have some amazing breakthroughs.

In the areas of human health, innovative technology, education and observations of Earth from space, there are already demonstrated benefits to people back on Earth. Lives have been saved, station-generated images assist with disaster relief, new materials improve products, and education programs inspire future scientists, engineers and space explorers. Some benefits in this updated second edition have expanded in scope. In other cases, new benefits have developed.

Since the publication of the first edition, a new constituency has developed, one that is using the ISS in a totally different fashion—to develop a commercial market in low-Earth orbit. From pharmaceutical companies conducting commercially-funded research on ISS, to private firms offering unique research capabilities and other services, to commercial cargo and crew, the ISS is proving itself to be just as adaptable to new business relationships as it has been for a broad diversity in research disciplines.

This book summarizes the scientific, technological and educational accomplishments of research on the space station that have had and will continue to have an impact to life on Earth. All serve as examples of the space station's potential as a groundbreaking research facility. Through advancing the state of scientific knowledge of our planet, looking after our health, developing advanced technologies and providing a space platform that inspires and educates the science and technology leaders of tomorrow, these benefits will drive the legacy of the space station as its research strengthens economies and enhances the quality of life here on Earth for all people.

Introduction

Welcome as we share the successes of the International Space Station (ISS) in this second edition of the *International Space Station Benefits for Humanity*. The ISS is a unique scientific platform that has existed since 1998 and has enabled over 2,400 researchers in 83 countries and areas to conduct more than 1,700 experiments in microgravity through just September 2014, and the research continues...

Since November 2, 2000, the ISS has maintained a continuous human presence in space. Even before it was habitable, the research began on the only orbiting laboratory of its kind. In 2011, when ISS assembly was complete, the focus shifted to fully utilizing the lab for continued scientific research, technology development, space exploration, commerce, and education.

The tremendous value of the ISS began through the engineering achievement evolving over a decade. Components were built in various countries around the world—all without the benefit of prior ground testing—allowing us to learn a vast amount about construction and about how humans and spacecraft systems function in orbit. This testament to the international achievement exemplifies cultural harmonization through cooperative teamwork leading to an international partnership that has continued to flourish and foster international cooperation. While each ISS partner has distinct agency goals for research conducted, a unified goal exists to extend the knowledge gleaned to benefit all humankind.

Value of the Platform

The Engineering Achievement

The International Achievement

The Research Achievement

In the first edition of the book released in 2012, the scientific, technological and educational accomplishments of ISS research that have an impact on life on Earth were summarized through a compilation of stories. The many benefits being realized were primarily in the areas of human health, Earth observations and disaster response, and global education.

This second edition includes updated statistics on the impacts of those benefits as well as new benefits that have developed since the first publication. In addition, two new sections have been added to the book: Economic Development of Space and Innovative Technology.

Economic Development of Space highlights case studies from public-private partnerships that are leading to a new economy in low-Earth orbit (LEO). Businesses provide both transportation to the ISS as well as some research facilities and services. These relationships promote a paradigm shift of government-funded, contractor-provided goods and services to commercially-provided goods purchased by government agencies. Other examples include commercial firms spending their research and development dollars to conduct investigations

Benefits of Research and Technology

on ISS and commercial service providers selling services directly to ISS users. This section provides examples of the use of ISS as a testbed for new business relationships and illustrates successful partnerships.

The second new section, Innovative Technology, merges technology demonstration and physical science findings that promise to return Earth benefits through continued research. Examples include robotic refueling concepts for life extensions of costly satellites in geo-synchronous orbit that have applications to the robotics industry on Earth, flame behavior experiments that reveal insight into how fuel burns in microgravity leading to the possibility of improving engine efficiency on Earth, and nanostructures and smart fluids examples of materials improvements that are being developed using data from ISS.

This publication also expands the benefits of research results in human health, environmental change and disaster response and in education activities developed to capture student imaginations in support of Science, Technology, Engineering and Mathematics, or STEM, education, internationally. Applications to human health of the knowledge gained on ISS continue to grow and improve healthcare technologies and our understanding of human physiology.

The ISS is a stepping stone for future space exploration, as the only orbiting multi-disciplinary laboratory of its kind returning research results that develop LEO and improve life on our planet. The goal of this publication is to serve as a source of pride to those who read it and learn of the unique shared laboratory orbiting our planet that provides ground for critical technologies and ways to keep humans healthy in space.

Benefits for Humanity Themes

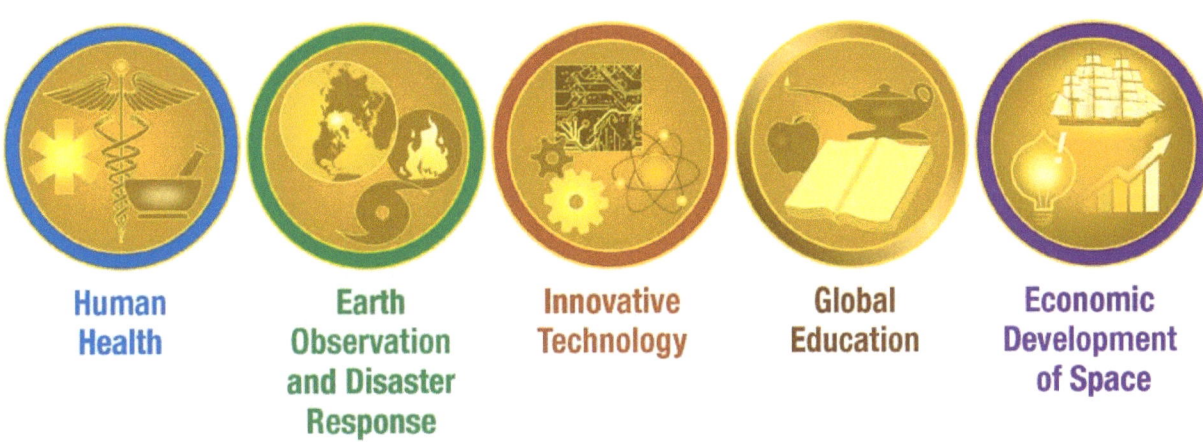

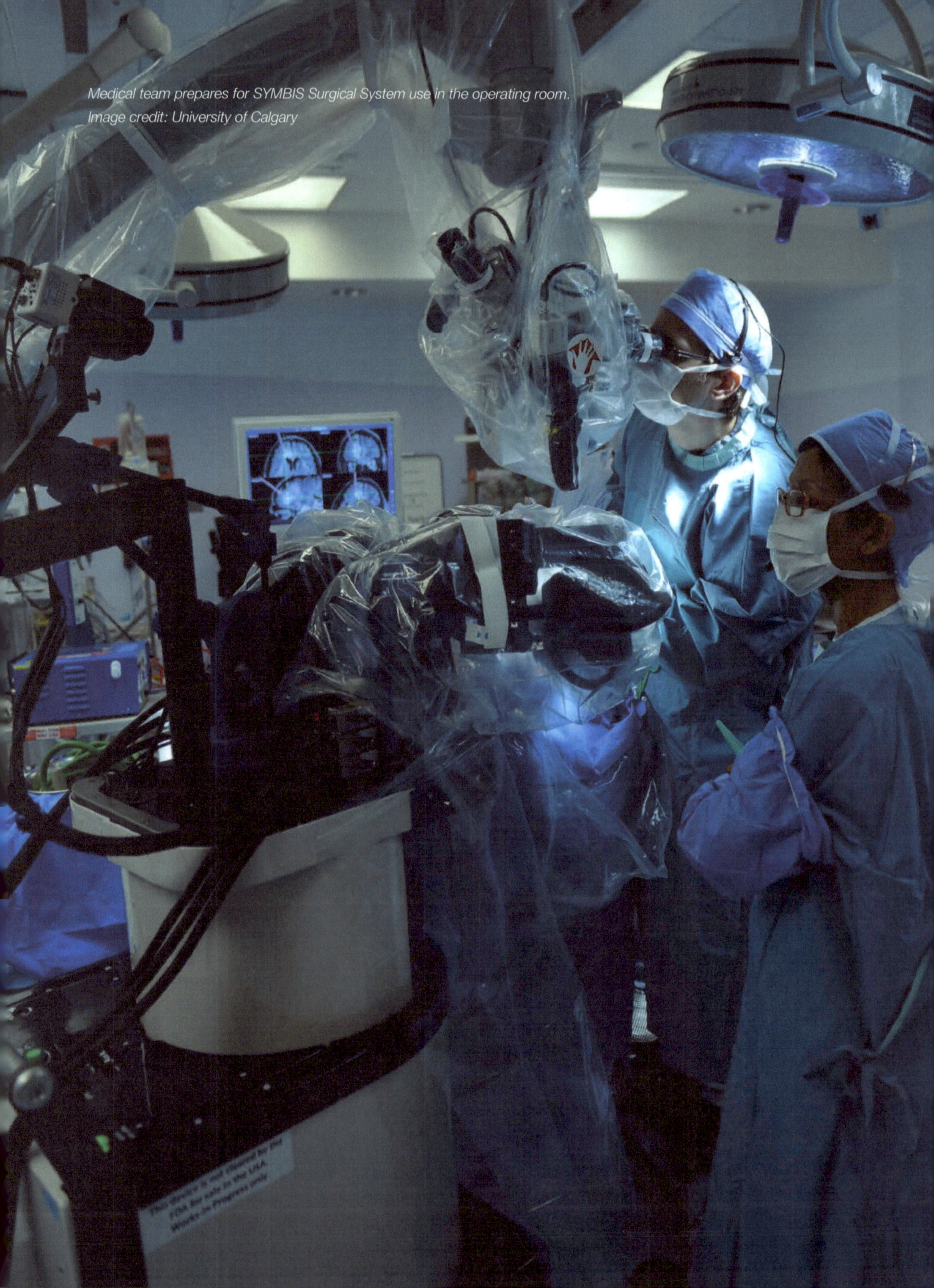
Medical team prepares for SYMBIS Surgical System use in the operating room.
Image credit: University of Calgary

Human Health

The International Space Station is a unique laboratory for performing investigations that affect human health both in space and on Earth. During its time in orbit, the space station has enabled research that is providing a better understanding of many aspects of human health including aging, trauma, disease and environmental impacts. Driven by the need to support astronaut health, several biological and human physiological investigations have yielded important results that we on Earth can also benefit from. These results include new ways to mitigate bone loss, insights into bacterial behavior, and innovative wound-healing techniques. Advances in telemedicine, disease models, psychological stress response systems, nutrition and cell behavior are just a few more examples of the benefits that have been gained from applying studies in orbit to human health back on Earth.

Health Technology

Research on ISS has allowed for innovations in surgical performance through the world's first robotic technology capable of performing surgery inside MRI machines. This technology is making difficult brain tumor surgeries easier and impossible surgeries possible. Soon, medical technology stemming from space station robotics will enter clinical trials for use in the early diagnosis and treatment of breast cancer by providing increased access, precision and dexterity resulting in highly accurate and minimally invasive procedures. Development of an advanced technology solution for pediatric surgery is also in the design stages. In common laser surgeries to correct eyesight, a new technology developed on ISS is now used on Earth to track the patient's eye and precisely direct a laser scalpel. Thermal regulation research on ISS has also led to the use of sensor technology for monitoring during surgery.

When medical facilities are not readily available such as in remote and underdeveloped regions of the world, ultrasound units are used in conjunction with protocols for performing complex procedures rapidly with remote expert guidance and training. These telemedicine and remote guidance techniques empower local healthcare providers, provide patients with access to more timely and diagnostic care, and the healthcare system is made more efficient.

A lightweight, easy-to-use device to measure nitric oxide in air exhaled by astronauts on ISS is used to study possible airway inflammation before health problems are encountered. This device is now used at some health centers to monitor levels of asthma control leading to more accurate medication dosing, reduced attacks, and improved quality of life.

The study of plasmas (charged gases that can permeate many materials and spread evenly and quickly) reveals that they support the disinfecting of chronic wounds, the neutralization of bacteria, the boosting of tumor inactivation, and even the jumpstarting plant growth.

Robotic arms lend a healing touch

The delicate touch that successfully removed an egg-shaped tumor from Paige Nickason's brain got a helping hand from a world-renowned arm—a robotic arm, that is. The technology that went into developing neuroArm, the world's first robot capable of performing surgery inside magnetic resonance machines, was born of the Canadarm (developed in collaboration with engineers a MacDonald, Dettwiler, and Associates, Ltd. [MDA] for the U.S. Space Shuttle Program) as well as Canadarm2 and Dextre, the Canadian Space Agency's family of space robots performing the heavy lifting and maintenance aboard the International Space Station.

neuroArm began with the search for a solution to a surgical dilemma: how to make difficult surgeries easier or impossible surgeries possible. MDA worked with a team led by Dr. Garnette Sutherland at the University of Calgary to develop a highly precise robotic arm that works in conjunction with the advanced imaging capabilities of magnetic resonance imaging (MRI) systems. Surgeons wanted to be able to perform surgeries while a patient was inside an MRI machine, which meant designing a robot that was as dexterous as the human hand but even more precise and tremor-free. Operating inside the MRI also meant it had to be made entirely from safe, MRI compatible materials (for instance, ceramic motors) so that it would not be affected by the MRI's magnetic field or, conversely, disrupt the MRI's images. The project team developed novel ways to control the robot's movements and give the robot's operator a sense of touch via an intuitive, haptic hand-controller located at a remote work station—essential so that the surgeon can precisely control the robot and can feel the tool-tissue interface during the surgery.

> Robotic specialists and surgeons sought to make difficult surgeries easier or impossible surgeries possible.

"Where the robot entered my head," says 21-year-old Paige Nickason, the first patient to have brain surgery performed by a robot, as she points to an area on her forehead. "Now that neuroArm has removed the tumor from my brain, it will go on to help many other people like me around the world."

Image credit: University of Calgary

Since Paige Nickason's surgery in 2008, neuroArm has been used in initial clinical experience with 35 patients who were otherwise inoperable. In 2010, the neuroArm technology was licensed to IMRIS Inc., a private, publicly traded medical device manufacturer based in Winnipeg, Manitoba, Canada, for development of the next-generation platform and for wide distribution under the name "SYMBIS Surgical System."

IMRIS is advancing the design to commercialize minimally invasive brain tumor resection procedures, which allow surgeons to see detailed, 3-D images of the brain as well as use surgical tools and hand controllers that allow the surgeon to feel tissue and apply pressure when he or she operates. SYMBIS has been undergoing calibration, testing and validation at Dr. Sutherland's research facility since March 2015. SYMBIS is expected to be able to perform microsurgery and stereotactic biopsy within the bore of the magnet while real-time MR images are being acquired. The system is more compact, with improved haptics, safety no-go zones, motion scaling and tremor filters. SYMBIS is currently being reviewed by the FDA, and once approved, the system will be made available commercially for other centers worldwide to establish its clinical efficacy through clinical trials.

MDA is also continuing to apply its space technologies and know-how to medical solutions for life on Earth. The company has partnered with the Hospital for Sick Children (SickKids) in Toronto, Ontario, to collaborate on the design and development of an advanced technology solution for pediatric surgery. Dubbed KidsArm, the sophisticated, teleoperated surgical system is being designed specifically to operate on small children and babies. KidsArm is intended for use by surgeons in conjunction with a high-precision, real-time imaging technology to reconnect delicate vessels such as veins, arteries or intestines.

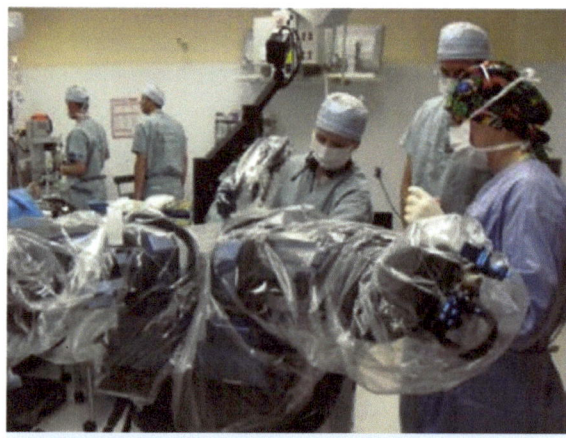

Medical team prepares for SYMBIS Surgical System use in the operating room.

Image credit: University of Calgary

In collaboration with the Centre for Surgical Invention and Innovation (CSII) in Hamilton, Ontario, MDA is also developing an advanced platform to provide a more accurate and less invasive identification and treatment of breast tumors in the MRI. The image-guided autonomous robot (IGAR) will provide increased access, precision and dexterity, resulting in more accurate and less invasive procedures. IGAR is currently in the second phase of clinical trials in Hamilton, Ontario, Canada, and Quebec City, Quebec, Canada.

 Watch these videos to learn more:

neuroArm: http://tinyurl.com/neuroArm

KidsArm: http://tinyurl.com/KidsArm

IGAR: http://tinyurl.com/CSA-IGAR

Robots from space lead to one-stop breast cancer diagnosis treatment

Technology derived from the highly capable robots designed for the International Space Station may soon increase access to life-saving surgical techniques to fight breast cancer.

> ISS technologies enable a robot to provide increased access, precision and dexterity, resulting in highly accurate and minimally invasive surgical procedures.

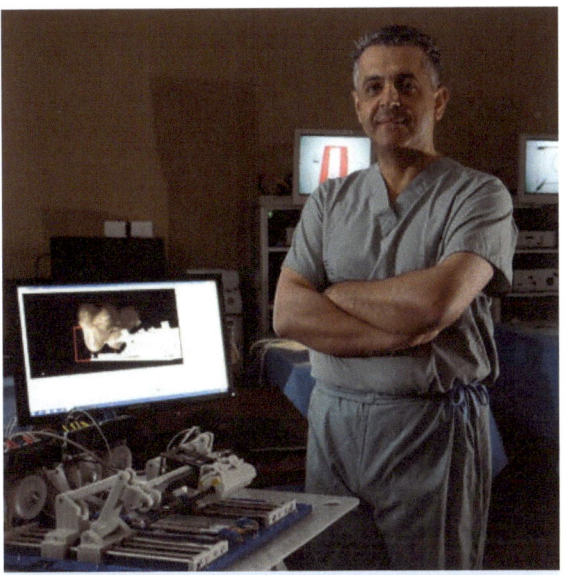

Dr. Mehran Anvari, chief executive officer and scientific director at the Centre for Surgical Invention and Innovation, with the Image-Guided Autonomous Robot (IGAR) manipulator.

Image credit: The Hamilton Spectator

A team of collaborative researchers with the Centre for Surgical Invention and Innovation (CSII) in Canada is working to enhance the quality and access to healthcare through the development and commercialization of innovative medical robotic technologies. In particular, an advanced platform is about to enter clinical trials for use in the early diagnosis and treatment of breast cancer.

The main player besides the medical staff is a robot. But not just any robot. This robot's technology was designed for use aboard the International Space Station by MacDonald, Dettwiler and Associates Ltd. (MDA) for the Canadian Space Agency (CSA).

Researchers created the Image-Guided Autonomous Robot (IGAR) from a long line of Canadian heavy lifters and maintenance performers for the space shuttle and space station Canadarm, Canadarm2 and Dextre. In dealing with breast cancer, IGAR is expected to provide increased access, precision and dexterity, resulting in highly accurate and minimally invasive procedures.

Dr. Mehran Anvari, chief executive officer and scientific director at CSII, said the IGAR platform moves the use of robotics in surgery to a new dimension, allowing the robot to act in an automated fashion after programming by a physician.

IGAR is designed to work in combination with an MRI scanner, which is highly sensitive to early detection of suspicious breast lesions before they possibly turn into a much larger problem. The radiologist uses specially designed software to tag the potential target and tell IGAR what path to take. The software then helps the radiologist to make sure he or she is accurately hitting the right area. IGAR has a special tool interface that can be used to define adaptors for any needle-based biopsy device or a wide range of instruments that remove tissue, known in the medical world as needle-based ablation devices.

Anvari explained that the automated robot is capable of placing the biopsy and ablation tools within 1 mm of the lesion in question with a high degree of targeting accuracy, improving sampling, reducing the pain of the procedure, reducing time in the MRI suite and reducing cost as a consequence. He also said that using the robot will allow all radiologists to perform this procedure equally well, regardless of the number of cases per year and move the site of treatment from operation room to radiology suite for a significant number of patients. The radiologist can operate in the challenging magnetic environment of the MRI, providing access to leading tumor-targeting technology. The robot fits on the patient bed, so it can travel in and out of the MRI opening easily. This in turn simplifies the flow of patients in the department, which

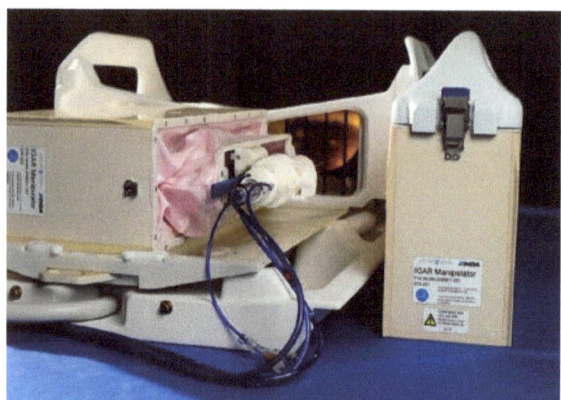

IGAR manipulator and full breast intervention platform mounted on the patient support structure with a biopsy tool attached.
Image credit: CSii and MDA

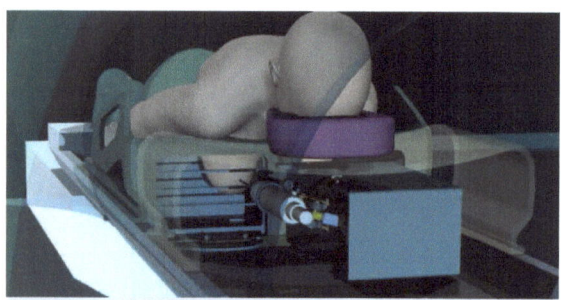

Artist rendering of IGAR performing a biopsy.
Image credit: CSii and MDA

can be challenging to many radiologists, optimizing patient time to diagnose.

Dr. Nathalie Duchesne, co-investigator on the clinical study and breast radiologist at the Saint-Sacrament Hospital in Quebec City, Quebec, Canada, has been teaching MRI-guided breast biopsy for years and will be performing the first of three clinical trials. She said there are many steps in the procedure that are operator-dependent, and these steps may prevent good sampling of the lesions if not done properly. Duchesne believes IGAR will decrease the time of the exam, ensure good sampling and increase patient's comfort during the exam. Duchesne and her team think that IGAR will improve sample collection because it will be less operator-dependent, and it will be constant from one doctor to another, from one patient to the other, and from one lesion to the other.

IGAR removes most of the "manual" aspects of the procedure and reduces user-dependence and the level of training required. This allows for a standard process regardless of experience. An expert will program remotely once the patient is in the MRI suite. A physician will then supervise to make sure the patient is comfortable and there are no complications.

Anvari said this technology lays the foundation for a family of telerobotic systems, and it has the potential to change the way people think about performing these interventions and ensures that specialized, highly-trained doctors are focusing on the activities to which their training is best suited. Anvari believes this technology will improve efficiency in the health care system by streamlining clinical workflow and allowing highly skilled radiologists to extend their care to a wider population through teleoperation.

This robotic technology is not limited only to biopsies. Duchesne explained that IGAR is paving the way for the minimally invasive excision and treatment of small tumors that are often found incidentally during pre-op MRI.

The trend toward breast preservation has brought on the importance of lumpectomies. For tumors that may require this procedure because they are invisible to ultrasound and X-ray mammography, researchers are currently developing the ability for IGAR to deploy a radioactive seed—smaller than a grain of rice—near the area of interest. During surgery, the seed can be located with a detector, allowing the doctor to identify the lesion and remove it with increased accuracy and patient comfort. It is expected that follow-up surgeries also will be greatly reduced.

Whether it be capturing a visiting spacecraft or helping save lives, Canadian-designed robots are lending a hand. Bringing beneficial technologies from the space station to the ground will hopefully one day allow us to make historic strides in cancer treatment.

 Watch this video to learn more about IGAR: http://tinyurl.com/CSA-IGAR

Improved eye surgery with space hardware

Laser surgery to correct eyesight is common practice, and technology developed for use in space is now commonly used on Earth to track the patient's eye and precisely direct the laser scalpel.

When looking at a fixed point while tilting or shaking one's head, a reflex allows the eyes to automatically

> The device developed for ISS allows the tracking of eye position without interfering with a surgeon's work during corrective laser eye surgery.

hold steady and see clearly even while this movement is taking place. This involves the brain constantly interpreting information from the inner ear to maintain balance and stable vision. An essential feature of this sensory system is the use of gravity as a reference.

The Eye Tracking Device experiment researched mechanisms involved in this process and how humans' frames of reference are altered in space. The experiment used a specially designed headset fitted with high-performance, image-processing chips able to track the eyes without interfering with an astronaut's normal work. The results showed that our balance and the overall control of eye movements are indeed affected by weightlessness. These two systems work closely together under normal gravity conditions but become somewhat dissociated in weightlessness.

After a flight, it takes several days to weeks for the astronauts to return to normal. The findings point to the entire sensory-motor complex and spatial perception relying on gravity as a reference for orientation.

In parallel with its use on the space station, the engineers realized the device had potential for applications on Earth. Tracking the eye's position without interfering with the surgeon's work is essential in laser surgery. The space technology proved ideal, and the Eye Tracking Device equipment is now being used in a large proportion of corrective laser surgeries throughout the world. A commercially available version has been delivered to a large number of research laboratories in Europe and North America for ground-based studies.

Sensor technologies for high-pressure jobs and operations

Novel sensor technologies used within the joint Thermolab experiment (2009-2012) of ESA/DLR have been used for improving our understanding of thermal regulation of astronauts in space. These sensor technologies also hold great potential and benefits for use within many different critical areas from fire-fighting to recognizing exhaustion or early overheating. In fact, the sensor is currently used in hospitals for monitoring during surgeries and on intensive care units.

Thermal regulation in the body is vital for our well-being. Our vital organs are kept at a constant temperature of 37° C (98.6 F) whether it is the middle of a freezing winter or on a hot sunny beach. Any disturbance to this stasis can cause symptoms such as physical and mental fatigue or, in the extreme, fatal effects on how the body functions under conditions such as heat stroke and hypothermia.

In weightlessness, the adaptation of the cardiovascular system, the lack of convection in space and the shifting of fluids to the upper half of the body could have a negative influence on thermal regulation.

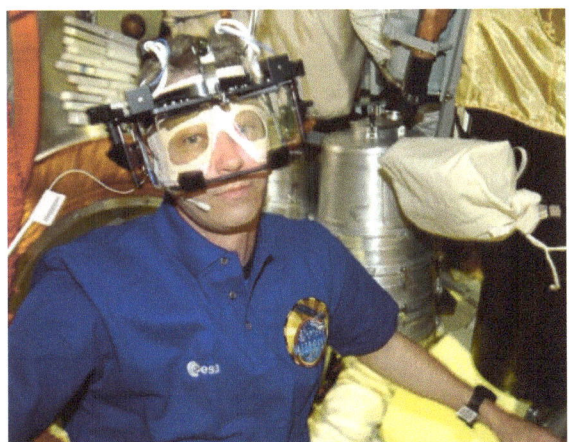

Former ESA astronaut Thomas Reiter undertakes the Eye Tracking Device experiment on the ISS in 2006.
Image credit: ESA

> Sensor technology developed on ISS is now used to monitor thermal regulation during surgeries and in intensive care units.

NASA astronaut Sunita Williams uses the Portable Pulmonary Function System whilst on the CEVIS cycle exercise device during a session of the joint Thermolab/EKE/VO2Max experiments in August 2012.

Image credit: NASA

The Thermolab experiment has been looking at changes in thermal regulation and cardiovascular adaptations in weightlessness by investigating how the body heats up and cools down during exercise. The testing of this new type of sensor to record the core body temperature in orbit could have novel applications in space and on Earth. This new sensor was developed for DLR by Charité (Berlin) and Draegerwerk (Lübeck) since standard ground measurement in clinics and surgeries use an internal body probe for taking measurements, which is not practical in orbit. The sensors measure the skin temperature and the heat flow in the skin, which are used to calculate core body temperature using sophisticated algorithms.

Compared to on Earth, core body temperature rises faster during exercise on the International Space Station. This is likely caused by fluid shifts and modified heat flow away from the body. It is also noticeable that the body temperature takes longer to cool down to core temperature after exercise. The measurement of the core body temperature together with cardiovascular measurements taken during NASA's VO2 Max protocol can be used to evaluate the subject's state of fatigue, which is very important during a space mission for optimising mission success. This makes this non-invasive double sensor a very useful diagnostic tool for recognising early warning signs of fatigue during spacewalks in orbit. On Earth, firefighters, jet pilots, miners, steel workers, soldiers in combat, divers, mountaineers, polar explorers, marine fishermen, and all who work in extreme conditions could benefit from the new measurement technology.

Bringing space station ultrasound to the ends of the Earth

Fast, efficient and readily available medical attention is key to survival in a health emergency. When a person is stricken with injury or illness, getting a quick and accurate diagnosis through medical imaging technology can be crucial for ensuring proper treatment. For people who live in major cities and towns where fully equipped hospitals are only a quick ambulance ride away, that's not usually a problem. But for those without medical facilities within easy reach, it can mean the difference between life and death.

For astronauts in orbit about 240 miles above Earth aboard the International Space Station, that problem was addressed through the Advanced Diagnostic Ultrasound in Microgravity (ADUM) investigation. Space station astronauts are trained to use a small ultrasound unit aboard the station to examine fellow crewmates. In the event of a health concern, astronauts could use this facility to diagnose many injuries and illnesses with the help of doctors on Earth. Launched in 2011, the ultrasound unit used for ADUM was replaced

> Medical care becomes more accessible in remote regions by use of small ultrasound units and tele-medicine, and remote guidance techniques, just like those on ISS.

with a smaller and even more sophisticated scanner dubbed Ultrasound 2, currently in use aboard the orbiting laboratory.

Now those same techniques are being adapted and used for people living in remote, underdeveloped areas where CT scans, MRIs and even simple X-ray exams are impossible. In partnership with the World Interactive Network Focused on Critical Ultrasound (WINFOCUS), ADUM principal investigator Scott Dulchavsky, M.D., is taking techniques originally developed for space station astronauts and adapting them for use in Earth's farthest corners by developing protocols for performing complex procedures rapidly with remote expert guidance and training.

WINFOCUS is a global network organization whose main goal is to use ultrasound as an enabling point-of-care device in an effort to make medical care more accessible in remote regions. Using the ADUM methods, WINFOCUS has trained over 20,000 physicians and physician extenders in 68 countries. This includes two important holistic healthcare projects: in remote areas of Nicaragua (from 2011) and in Brazil in a statewide healthcare project in partnership with the Secretary of Health of the State of Minas Gerais (since 2012).

WINFOCUS has also benefited from the tele-medicine and remote guidance techniques developed for use on the space station, and has adapted and further developed them in order to allow large-scale integration in healthcare systems on Earth through low-cost applications. Local healthcare providers are empowered, more patients can access quality and timely diagnostic care, and the healthcare system is made more accessible and efficient.

ADUM's impact is also felt in modern emergency rooms, proving the effectiveness of ultrasound in diagnosing conditions previously considered beyond its technical capabilities, such as a collapsed lung, which has now become integrated as a standard of care in medical treatments. In addition, the ADUM protocols have proven so effective that they're now part of the standard medical school curriculum. The American College of Surgeons, which requires ultrasound training for all surgical interns and residents, is using the ADUM program.

The ADUM investigation and the WINFOCUS partnership have brought the promise of space station research back down to Earth in perhaps the most direct and immediate way possible—keeping people healthy and alive, even in remote regions where care was previously a limited option.

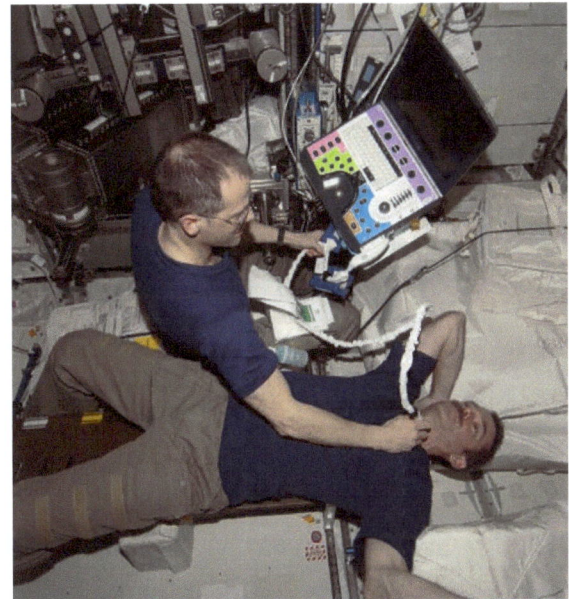

NASA astronaut Tom Marshburn assists Canadian Space Agency astronaut Chris Hadfield with an Ultrasound 2 scan in the Columbus Module of the International Space Station.

Image credit: NASA

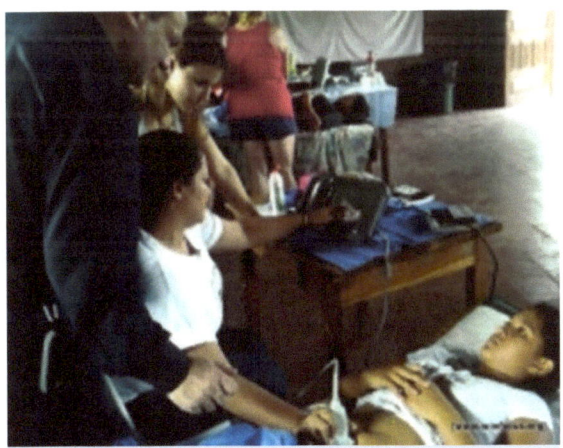

World Interactive Network Focused on Critical Ultrasound (WINFOCUS) and Henry Ford Innovation Institute members, Dr. Luca Neri and Alberta Spreafico work with Kathleen Garcia from Wyle Engineering to help train Dr. Chamorro from the rural community of Las Salinas, Nicaragua, using the Advanced Diagnostic Ultrasound in Microgravity and tele-ultrasound applications.

Image credit: WINFOCUS/Missions of Grace

Are you asthmatic? Your new helper comes from space

Kalle, a 10-year-old boy, is already in favor of space technology. In the future, he could control his asthma with a small device also used by crew members aboard the International Space Station. Because of it, he knows almost everything about nitric oxide—an important gas we all breathe out.

Nitric oxide, or nitrogen monoxide, as it is properly called, is both a good and bad molecule, found almost everywhere as an air pollutant that is produced by vehicle exhaust and industrial processes burning fuel. Nitric oxide is a contributor to the damage of the ozone layer and easily converts into nitric acid—which may fall as acid rain.

Intriguingly, tiny amounts of nitric oxide are released locally in inflamed tissue of humans and other mammals. Tracing it back to its source can reveal different diseases.

> A lightweight, easy-to-use device monitors levels of asthma control leading to more accurate medication dosing, reduced attacks, and improved quality of life.

In people with asthma, inflammation in the lung adds nitric oxide to exhaled air. Measuring the gas can help to diagnose the disease and may prevent attacks if the levels of nitric oxide indicate that medication should be adjusted.

Nitric oxide is also an interesting molecule on the space station. Dust and small particles floating around in weightlessness can be inhaled by the astronauts, possibly triggering inflammation of the airways. It also plays a role in decompression sickness that may arise from spacewalks.

The European Space Agency (ESA) uses a lightweight, easy-to-use, accurate device for measuring nitric oxide in exhaled air. The aim is to investigate possible airway inflammation in astronauts and act before it becomes a health problem.

Following its development by the Swedish company Aerocrine AB and ESA, the device has been found beneficial in space exploration and everyday use on Earth.

NIOX MINO® is now used by patients like Kalle at health centers. They can monitor levels of asthma control and the efficiency of medication—leading to more accurate dosing, reduced attacks and improved quality of life.

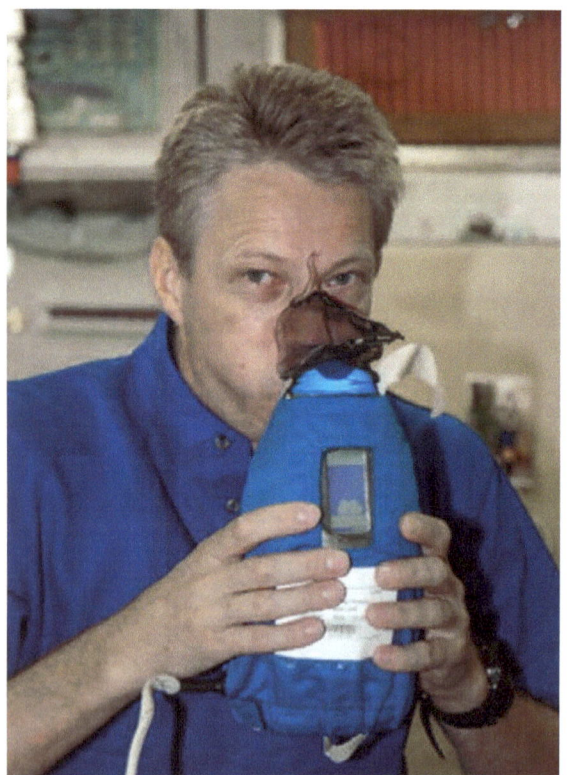

Former European Space Agency (ESA) astronaut Thomas Reiter undertakes science activities for the Nitric Oxide Analyzer experiment in 2006.
Image credit: ESA

Cold plasmas assist in wound healing

A unique form of matter could help disinfect wounds, neutralize bacteria, help people heal faster, and even fight cancer—and its potential for human health is now well understood, thanks to research on the International Space Station. The microgravity environment provides a powerful method for studying plasmas, one of the four states of matter along with liquid, solid and gas.

The Plasma Kristall Experiment (PK-3 Plus) lab, a Russian-German collaboration, provided new insight into an unusual type of matter known as plasma crystals.

> Plasma studies reveal applications to disinfect chronic wounds, neutralize bacteria, boost tumor inactivation, and jumpstart plant growth.

Russian cosmonaut Oleg Kotov, Expedition 30 flight engineer, inspects the Plasma Kristall Experiment laboratory, enclosed in black housing, in its new home in the Poisk Mini-Research Module 2 of the International Space Station.

Image credit: RKK-Energia

Because it is a charged gas, plasma can permeate many materials, spreading evenly and quickly. It can disinfect surfaces, and has been proven to neutralize drug-resistant bacteria like methicillin-resistant *Staphylococcus aureus* within seconds. In more than 3,500 examples in several clinical trials, physicians found plasmas can disinfect chronic wounds and help wounds heal faster. Other research has shown that along with chemotherapy, plasma treatment efficiently fights cancer; it can boost tumor inactivation by 500 percent, compared with just chemotherapy. Plasmas can even jumpstart plant growth.

For the researchers involved in PK-3, the technical challenges of space-based research provided the knowledge base for the medical spin-offs, according to Professor Gregor E. Morfill, director at the Max Planck Institute for Extraterrestrial Physics in Garching, Germany. Without space station research, some team members would never have been involved in plasma medicine.

The PK-3 lab was designed to study complex or "dusty" plasmas, which get their name from the presence of small, solid particles mixed into the plasma's charged gases. These particles can dramatically change the behavior of a plasma, and sometimes the particles even form crystalline structures. Dusty plasmas are found near artificial satellites, occur in Earth's upper atmosphere, and can be produced in lab settings. Physicists favor them because they are relatively easy to control and provide a unique view of physics at the single-particle level. But they can be difficult to study on Earth, because the planet's gravity affects the way dust particles settle and how they crystallize. This isn't the case on the space station, however.

Investigations with PK-3 Plus created dusty plasmas containing argon or neon gas as well as micron-size particles. The gas molecules received an electric charge so they would ionize and form a plasma, and then particles were injected into it. A laser lit up the sample while a camera recorded the particles moving through the plasma and organizing themselves in

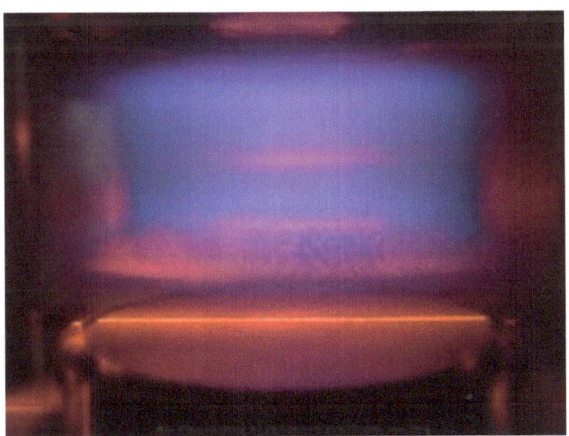

Side view of a plasma crystal in the laboratory. Dust particles are suspended in an argon plasma above a high-frequency electrode (bottom). The horizontal field of view is 2 cm.

Image credit: Max Planck Institute for Extraterrestrial Physics

crystal structures. Basic experiments tested a wide range of particle sizes and different gas types, and researchers found a plethora of interesting new phenomena. In one example, researchers used the PK-3 Plus high-resolution camera to examine the exact point at which matter changes its phase from liquid to solid. Other experiments tested how radio waves cause particles in a dusty plasma to move.

Beyond basic science, dusty plasmas have several practical applications in space and on Earth. For instance, some computer chips are manufactured using a processing plasma, and removal of microscopic particles is crucial for preventing chip contamination. Understanding how gases and dusty plasmas interact is critical for improving this technology. A better grasp of this interaction could also help scientists create powders containing specific ingredients, for applications like agriculture, hygiene and medicine. And plasmas hold great promise for treating sick and injured people on Earth.

Astronauts and cosmonauts operated the PK-3 Plus equipment during 20 separate missions across a six-year period, each lasting about five days. All told, collaborators on the PK-3 Plus investigation and its predecessor, PKE-Nefedov, have published more than 70 scientific papers and given at least 100 presentations at scientific conferences.

The past PK investigations may be concluded in space, but plasma medicine research in particular continues to produce new applications—which will further increase with the PK-4 investigation for which new hardware was commissioned on ISS in November 2014.

Preventing Bone Loss

The common problem of bone loss in the elderly is also observed in astronauts when they are in space. Ongoing studies on ISS indicate a reduction in bone loss and renal stone risk through use of a bisphosphonate and exercise to increase bone load and muscle training, and in a well-balanced, low-sodium diet. In promoting the health of the elderly at risk of osteoporosis, improved scanning technologies are under development to provide a reference technique to enable the early detection of osteoporosis and in the development of more effective countermeasures to its effects.

Preventing bone loss in spaceflight with prophylactic use of bisphosphonate: Health promotion of the elderly by space medicine technologies

Bone loss and kidney stones are well-known as essential problems for astronauts to overcome during extended stays in space. Crew members engage in physical exercise for two-and-a-half hours a day, six times a week (15 hours a week) while in orbit to avoid these issues. Nevertheless, the risks of these problems occurring cannot be completely eliminated through physical exercise alone.

> Ongoing studies indicate a reduction in bone loss and renal stone risk through use of a bisphosphonate and exercise.

JAXA astronaut Soichi Noguchi performs exercise aboard the International Space Station.
Image credit: JAXA/NASA

Bone plays an important role as a structure that supports the body and stores calcium. It retains fracture resistance by remodeling through a balance of bone resorption and formation. In a microgravity environment, because of reduced loading stimuli, there is increased bone resorption and no change in or possibly decreased bone formation, leading to bone mass loss at a rate of about 10 times that of osteoporosis. The proximal femoral bone loses 1.0 to 1.5 percent of its mass per month, or roughly 6 to 10 percent over a six-month stay in space, with the recovery after returning to Earth taking at least three or four years. The calcium balance (the difference between intake and excretion), which is about zero on Earth, decreases to about -250 mg/day during flight, a value that increases the risk of kidney stones.

Bisphosphonate is a therapeutic agent that has been used to treat osteoporosis patients for more than

Astronauts enjoy meals aboard the International Space Station.
Image credit: JAXA/NASA

a decade, with a proven efficacy to increase bone mass and decrease the occurrence of bone fracture. Through 90-day bed rest research on Earth, we confirmed that this agent has a preventive effect on the loss of bone mass. Based on these results as well as studies conducted by others, Japan Aerospace Exploration Agency (JAXA) and NASA decided to collaborate on a space biomedical experiment to prevent bone loss during spaceflight. Dr. Adrain Leblanc, United Space Research Association, and Dr. Toshio Matsumoto, Tokushima University, are the two principal investigators of this study.

JAXA and NASA crew members are participating in this study by taking this agent once a week while in space. The study is still ongoing; however, early results suggest that astronauts can significantly reduce the risk of bone loss and renal stones with the combination of resistive exercise and an antiresorptive such as a bisphosphonate.

Bone loss is also observed in bedridden older people. Elderly people lose 1 or 2 percent per year of their bone mass because of aging and a decline in the amount of female hormone. Osteoporosis is declared when a person has a bone mass 30 percent lower than the average for young adults, which is a condition affecting 13 million Japanese and one in two women aged 70 years and older. Every year, 160 thousand patients undergo operations for femoral neck fractures in Japan, followed by intense rehabilitation for three months. Such operations cost 1.5 million yen per person, and the total annual expense for medical treatments and care of these bone fractures amounts to 66.57 billion yen in total national cost.

The three key elements for promoting the health of elderly people to prevent fractures are nutrition, exercise and medicine. Meals should be nutritionally balanced with calcium-rich foods (milk, small fish, etc.) and vitamin D (fish, mushrooms, etc.). Limited sunbathing is also important for activation of vitamin D. Physical exercise to increase bone load and muscle training should also be integrated into each person's daily life. Those at high risk for fractures should take effective medicines to reduce the risk of fractures.

Accordingly, the secrets of the promotion of astronauts' health obtained from space medicine are expected to be utilized to promote the health of elderly people and the education of children.

Improved scanning technologies and insights into osteoporosis

ESA's Early Detection of Osteoporosis in Space (EDOS) experiment has been testing skeletal adaptation to long-term space exposure by using 3-D peripheral quantitative computed tomography (3DpQCT) as a technique for detection of bone structure. It has been providing a detailed evaluation of the bone loss and of kinetics of recovery after flight. ESA supported the development of the enhanced 3-D scanner by the Institute for Biomedical Engineering in Zürich and Scanco Medical as part of ESA's Microgravity Applications Programme (MAP). The scanner is providing high-quality, 3-D images of living bone structures as part of this ground experiment. This is backed up by analysis of bone biochemical markers in blood samples.

One important element that has derived from this research into bone loss in space is the successful commercialisation of the 3DpQCT scanner, of which ESA supported the development, for a non-invasive/in vivo technique for observation of bone structure.

The EDOS project has been assessing the efficiency of such a technique and will contribute to the development of a reference technique to perform an early detection of osteoporosis on Earth in a unique way. These improved diagnostics in the early stages of such a medical condition may prove extremely important in development of more effective countermeasures to the effects of osteoporosis. In 2006, according to the International Osteoporosis Foundation, 8.9 million fractures were estimated worldwide. The project will continue within the EDOS-2 project, which will commence in collaboration with Russia in spring 2015 in conjunction with the first one-year mission.

> Early detection of osteoporosis, and the development of more effective treatments, link astronauts to patients on Earth.

Xtreme CT distal radius.
Image credit: SCANCO Medical

Good diet, proper exercise help protect astronauts' bones

Eating right and exercising hard in space helps protect International Space Station astronauts' bones, a finding that may help solve one of the key problems facing future explorers heading beyond low-Earth orbit.

A study published in the September 2012 issue of the *Journal of Bone and Mineral Research* looked at the mineral density of specific bones as well as the entire skeleton of astronauts who used a new, stronger "weight lifting" machine. Of course, weights don't really

> After 51 years of human spaceflight, we have made significant progress in protecting bone health through diet and exercise.

"weigh" anything on the space station, but resistance machines allow astronauts to get the same kind of workout. The new Advanced Resistive Exercise Device (ARED), installed in 2008, doubles the maximum simulated weight to as much as 600 pounds.

Researchers compared measurements from 2006 until 2008 when astronauts used a less capable workout machine. They found that astronauts using the advanced system came home with more lean muscle and less fat and kept more of their whole body and regional bone mineral density. Those same astronauts also consumed sufficient calories and vitamin D, among other nutrients. These factors are known to support bone health and likely played a contributing role.

After 51 years of human spaceflight, these data mark the first significant progress in protecting bone through diet and exercise. Since the 1990s, resistance exercise has been thought to be a key method of protecting astronauts' bones. Normal, healthy bone constantly breaks down and renews itself, a process called remodeling. As long as these processes are in balance, bone mass and density stay the same. Earlier studies of Russian Mir space station residents found an increased rate of breakdown but little change in the rate of regrowth, resulting in an overall loss in bone density.

In the new study, astronauts who used the ARED device still had increased bone breakdown, but their bone renewal tended to increase, likely resulting in a better balance in whole bone-mineral density.

Bone density loss in astronauts on long-duration missions has been a major medical concern. In the past, astronauts have lost an average of 1 to 2 percent per month. By comparison, an elderly person loses about 1 to 2 percent per year.

This study shows that, through proper exercise and nutrition, crew members on long journeys in space can return to Earth with much less loss of bone mineral density. But a key question remains as to whether the bones are as strong as when the astronaut launched into space. For these and other reasons, additional studies to evaluate bone strength before and after flight are currently under way.

Beyond bone strength, further study is needed to figure out the best possible combination of exercise and diet for long-duration crews. One experiment on the space station right now is looking at how different ratios of animal protein and potassium in the diet affect bone health. Another is looking at the benefits

NASA astronaut Jeffrey Williams, Expedition 22 commander, exercises using the Advanced Resistive Exercise Device in the Tranquility node of the International Space Station.

Image credit: NASA

NASA Astronaut Dan Burbank, Expedition 30 commander, exercises using the Advanced Resistive Exercise Device aboard the International Space Station.

Image credit: JAXA/NASA

of lowering sodium intake. NASA food scientists have reformulated more than 80 space foods to reduce the sodium content.

Information gained through space station studies like these will be critical in enabling humans to explore destinations beyond low-Earth orbit.

Add salt? Astronauts' bones say please don't

Osteoporosis is a harsh disease that reduces the quality of life for millions and costs Europe around €25 billion ($31 billion) each year. It typically affects the elderly, so the rise in life expectancy in developed countries means the problems inflicted by osteoporosis are increasing.

Fortunately, research done in space may change the game. Astronauts on the International Space Station experience accelerated osteoporosis because of weightlessness, but it is carefully controlled, and they can regain their lost bone mass in time once they are back on Earth.

Studying what happens during long spaceflights offers a good insight into the process of osteoporosis—losing calcium and changing bone structure—and helps to develop methods to combat it.

It has been known since the 1990s that the human body holds on to sodium, without the corresponding water retention, during long stays in space. But the

> ISS research provides insight into the benefits of reduced sodium and increased bicarbonate consumption for those prone to osteoporosis.

European Space Agency (ESA) astronaut André Kuipers (left) and Russian cosmonaut Oleg Kononenko (right) with food items on the International Space Station in December 2011. In the SOdium LOad in microgravity experiment, astronaut subjects undergo two different diet regimes to determine the physiological effects of sodium on the body.
Image credit: ESA

textbooks said this was not possible. "Sodium retention in space" became an important subject to study.

Salt intake was investigated in a series of studies, in ground-based simulations and in space, and it was found that not only is sodium retained (probably in the skin), but it also affects the acid balance of the body and bone metabolism. So, high salt intake increases acidity in the body, which can accelerate bone loss.

The European Space Agency's (ESA's) recent SOdium LOad in microgravity (SOLO) study zoomed in on this question.

Nine crew members, including ESA's Frank De Winne and Paolo Nespoli during their long-duration flights in 2010 and 2011, followed low- and high-salt diets. The expected results may show that additional negative effects can be avoided either by reducing sodium intake or by using a simple alkalizing agent like bicarbonate to counter the acid imbalance.

This space research directly benefits everybody on Earth who is prone to osteoporosis.

European Space Agence (ESA) astronaut Frank De Winne undertakes a body mass measurement, an essential element of the SOdium LOad in microgravity experiment, on the space station.
Image credit: ESA

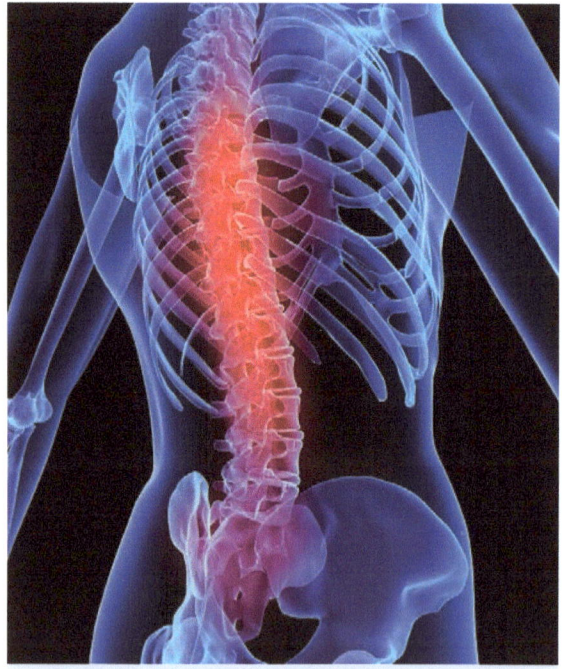

The SOdium LOad in microgravity experiment carries out research into salt retention and its effect on bone metabolism in astronauts, which can help provide insights into medical conditions on Earth, such as osteoporosis.
Image credit: Istockphoto/S.Kaulitzki

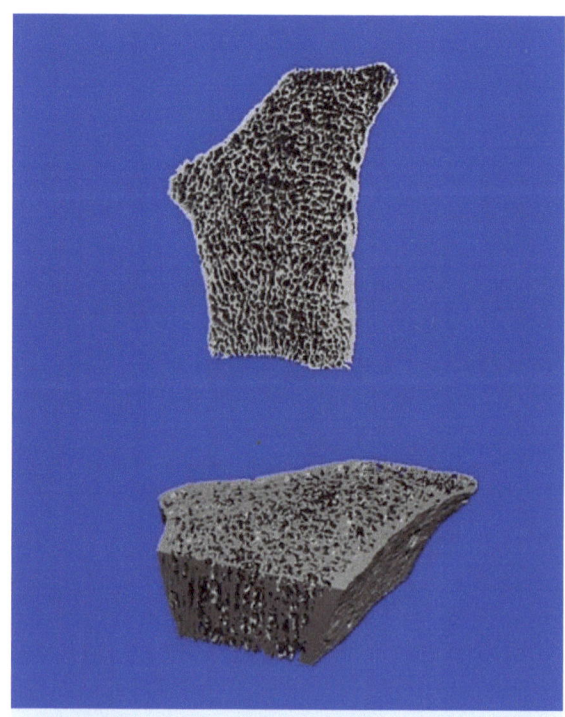

3-D pQCT image of osteoporotic bone.
Image credit: Scanco Medical AG

Immune Defenses

Virtually the entire population is infected with one of eight herpes viruses, four of which reactivate and appear in body fluids in response to the stress of spaceflight. A patent-pending device designed for use in either a doctor's office or on a spacecraft allow for the rapid detection of one of these viruses (VZV), which can lead to earlier treatment and prevent the onset of painful shingles. Microgravity studies on ISS help researchers pinpoint genetic triggers for immune responses in T-cells leading to future medical treatments on Earth for immunosuppression. Determining the changes that occur to the immune system in space is providing the means to develop targeted countermeasures to adverse effects in space, as well as providing additional information for targeted treatments on Earth for the development of pharmaceuticals that can suppress immune response to help manage autoimmune diseases or organ transplants.

Early detection of immune changes prevents painful shingles in astronauts and in Earth-bound patients[1]

The physiological, emotional and psychological stress associated with spaceflight can result in decreased immunity that reactivates the virus that causes shingles, a disease punctuated by painful skin lesions. NASA has developed a technology that can detect immune changes early enough to begin treatment before painful lesions appear in astronauts and people here on Earth. This early detection and treatment will reduce the duration of the disease and the incidence of long-term consequences.

> Space research has led to the rapid detection of Varicella (chickenpox virus), which improves treatment of shingles.

Spaceflight alters some elements of the human immune system: innate immunity, an early line of defense against infectious agents, and specific components of cellular immunity are decreased in astronauts. Astronauts do not experience increased incidence or severity of infectious disease during short-duration spaceflight, but NASA scientists are concerned about

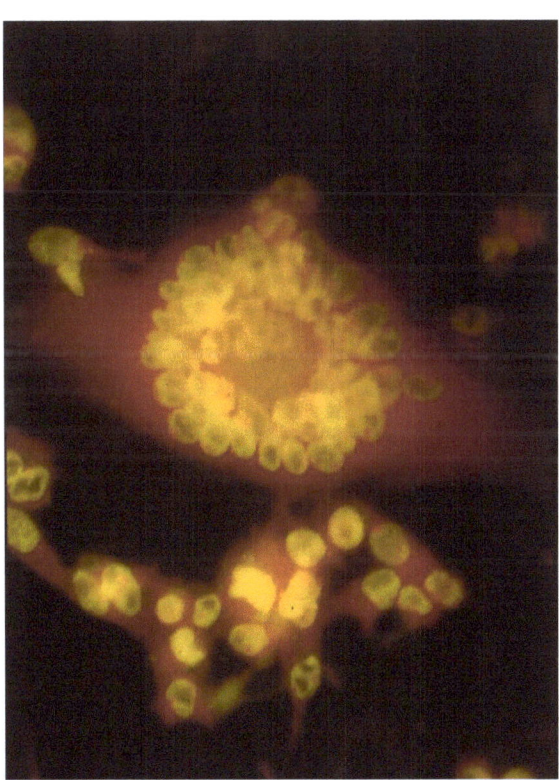

Varicella zoster-infected MeWo cells showing typical herpes virus-induced, multinucleated giant cells. Cultures are stained with acrydine orange to identify RNA (red) in the cytoplasm.

Image credit: NASA

---

[1]Adapted from an original article that appeared in NASA Technology Innovation, Vol 15; 3, 2010; NP-2010-06-658-HQ.

how the immune system will function over the long stays in space that may be required for exploration missions.

To determine specific causes of decreased immunity in healthy individuals is difficult, but the herpes viruses have become valuable tools in early detection of changes in the immune system, based largely on the astronaut studies. Eight herpes viruses may reside in the human body, and virtually all of us are infected by one or more of these viruses. Herpes viruses cause diseases including common "fever blisters" (herpes simplex virus or HSV), infectious mononucleosis (Epstein-Barr virus or EBV), chickenpox and shingles (varicella zoster virus or VZV). In immune-suppressed individuals, herpes viruses may cause several types of cancer, such as carcinoma, lymphoproliferative disease and others.

According to the Centers for Disease Control and Prevention, one million cases of shingles occur yearly in the U.S., and 100,000 to 200,000 of these cases develop into a particularly painful and sometimes debilitating condition known as post-herpetic neuralgia, which can last for months or years. The other seven herpes viruses also exist in an inactive state in different body tissues much like VZV, and similarly they may also reactivate and cause disease during periods of decreased immunity.

The most common cause of decreasing immunity is age, but chronic stress also results in decreased immunity and increases risk of the secondary disease, such as VZV-driven shingles. Chemotherapy, organ transplants and infectious diseases, such as human immunodeficiency virus (HIV), also result in decreased immunity. Thus, viral reactivation has been identified as an important indicator of clinically relevant immune changes. Studies of immune-compromised individuals indicate that these patients shed EBV in saliva at rates 90-fold higher than found in healthy individuals.

The herpes viruses are already present in astronauts, as they are in at least 95 percent of the general adult population worldwide. So measuring the appearance of herpes viruses in astronaut body fluids is critical. It is widely believed that various stressors associated with spaceflight are responsible for the observed decreased immunity. Researchers at NASA's Johnson Space Center found that four human herpes viruses reactivate and appear in body fluids in response to spaceflight. Due to the reduced cellular immunity, the viruses can emerge from their latent state into active infectious agents. The multiplying viruses are released into saliva, urine or blood and can be detected and quantified by a method called polymerase chain reaction (PCR) for each specific virus. The finding of VZV in saliva of astronauts was the first report of VZV being reactivated and shed in asymptomatic individuals, therefore posing a risk of disease in uninfected individuals. However, the PCR assay requires large, complex equipment, which is not practical for spaceflight.

To overcome this obstacle, NASA developed a rapid method of detection of VZV in body fluids, and a patent application is currently pending for it. The new technology requires a small sample of saliva, which is mixed with specialized reagents that produce a red color only when VZV is present. This technology makes possible early detection, before the appearance of skin lesions. Early detection allows for early administration of antiviral therapy and thus limits nerve damage and prevents overt disease. The device is designed for use in doctors' offices or spacecraft and can be modified easily for use with other viruses in saliva, urine, blood and spinal fluid. The sensitivity and specificity emanates from an antibody-antigen reaction.

In another collaborative study, NASA and University of Colorado Health Science Center (Denver) researchers developed a tool to assess stress hormones during space shuttle missions. Saliva samples are collected on individual filter paper strips and tested once back on Earth. The test measures cortisol and dehydroepiandrosterone (DHEA), two important stress and immune regulatory hormones. The filter paper also can be used for proteins and other molecules of interest in saliva. Booklets of these filter papers now are being used in university and government laboratories for remote saliva collection. These studies demonstrate the potential value of bringing to the general public a technology that could prevent a painful and debilitating condition in up to one million people each year in the U.S. alone.

Station immunology insights for Earth and space

When people get sick, their immune systems kick into gear to tell their bodies how to heal. T-cells—white blood cells that act like tiny generals—order an army of immune cells to organize and attack the enemy. Microgravity studies aboard the International Space Station are helping researchers pinpoint what drives these responses, leading to future medical treatments on Earth.

Scientists have known since the early days of human spaceflight that living in microgravity suppresses the immune system. During the Apollo Program, for instance, 15 of the 29 astronauts developed an

> Microgravity studies on ISS help researchers pinpoint genetic triggers for immune responses in T-cells leading to future medical treatments.

infection either during or right after flight. Forty years later, Leukin results show that immunosuppression begins within the first 60 hours of flight.

Findings from this investigation, led by Millie Hughes-Fulford, Ph.D., a former NASA astronaut and director of the Laboratory of Cell Growth at the University of California, San Francisco, enabled researchers to pinpoint some specific genetic triggers for the go/no go of the immune system responses in the T-cells. It was the first time scientists have been able to prove that gravity is making a difference in activation of the T-cell. A healthy body depends on these T-cells giving orders for the immune system to function properly as it marches into battle. There are factors that can hinder victory, however, such as signal interruption, delayed responses or even outright cell death. A suppressed immune system is like an army with an ineffective leader, significantly reducing the chances of a successful fight.

Expedition 30 Flight Engineer Andre Kuipers, European Space Agency, works with the Kubik facility in the Columbus Module of the International Space Station.
Image credit: NASA

Results revealed that specific genes within T-cells showed down regulation—a decrease in cell response—when exposed to microgravity. This combined down regulation in the genetics of T-cells leads to a reduction in the body's defense against infections during spaceflight in various ways. For instance, there is a reduced proinflammatory response, the cell's protective reaction to initiate healing. Cells also produce fewer cytokines, the proteins responsible for signaling communications between cells. There is even a negative impact to a cell's ability to multiply, known as mitogenesis, the chromosomal splitting in a cell nucleus necessary for cell reproduction.

Examples of immunosuppression on Earth include the AIDS-related HIV infection, rheumatoid arthritis and even age-related impacts to the immune system, which is why the elderly have a difficult time fighting off infections like pneumonia. Identifying how the immune system works at the cellular level provides a powerful tool to develop treatments at the root of the defense response. This is like a negotiation for peace talks before conflict breaks out, instead of trying to raise a white flag in the midst of an already raging battle. If doctors can isolate and control specific immune responses, they increase the chance for recovery. With the removal of gravity as its own variable, the data gathered from immune studies in space can be used to help understand some of the immune challenges seen in these populations on Earth.

Hughes-Fulford launched a follow-on immunology study aboard the space station, funded by a grant from the National Institutes of Health and sponsored by the Center for Advancement of Science in Space (http://www.iss-casis.org/). Launching on the SpaceX-3 commercial resupply mission, the investigation, called T-Cell Activation in Aging, investigates at another class of control points in T-cells that trigger immune response. Finding the genes that tell the cells to turn on and off is key to advancing medical options to improve immune system functions. Data analysis is underway, with the potential to pinpoint new candidate pharmaceutical targets to treat immunosuppression.

Targeted treatments to improve immune response

Cell biology experiments have been uncovering different aspects of altered immune system response in weightlessness. Determining the changes that occur to the immune system in space is providing the means to develop targeted countermeasures to adverse effects in space, as well as providing

> ISS research provides the means to develop pharmaceuticals and targeted treatments that can suppress immune response to help deal with autoimmune diseases or organ transplants.

additional information for targeted treatments on Earth. This could either be for the purpose of developing pharmaceuticals that can improve treatment and recovery from certain medical conditions or alternatively targeted treatments that can suppress immune response, for example to help deal with autoimmune diseases or organ transplants.

Research undertaken in the Kubik incubators has uncovered many altered mechanisms that occur in the immune system in space using biological samples processed at body temperature at 0 g and 1 g (centrifuge) in orbit. This has included discovering reduced function in monocyte white blood cells that is due to a disrupted cytoskeleton. This is an apparent inhibition of the Protein Kinase C family of enzymes and a specific immune cell transmitter, called the Rel/NF-κB pathway, which stops working in weightlessness. All of these are important mechanisms in immune response.

One of the most recent ESA experiments in this domain was the ROle of Apoptosis in Lymphocyte Depression (ROALD) experiment series, which was undertaken in 2008 with a follow-up experiment in 2011. In the first part of the experiment, researchers discovered that a particular enzyme called 5-LOX, which in part regulates the life expectancy of human cells, became more active in weightlessness and could play a real role in causing weakened immune systems. The 5-LOX enzyme can be blocked with existing drugs, so using these findings to improve human health could be a close reality. Additional efforts to understand this treatment pathway, targeting patient treatment on Earth, is ongoing.

ESA astronaut Thomas Reiter undertakes in-orbit activities for one of ESA's immunology experiments in 2006.
Image credit: ESA

Developing New Therapies

Studying the unique and complicated structures of proteins in the human body leads to the development of medical treatments. Microgravity allows unique conditions for growth of protein crystals where there is no gravity or convection to disrupt their growth. The protein expressed in certain muscle fibers of patients with Duchenne Muscular Dystrophy, which affects 1 in 3,500 boys, has been successfully crystallized in space revealing a new inhibitor several hundred times stronger than the prototype inhibitor.

Microencapsulation is the process by which tiny, liquid-filled, biodegradable micro-balloons are created containing specific combinations of concentrated anti-tumor drugs. The goal is to deliver this medication using specialized needles to specific treatment sites within a cancer patient. The microgravity environment, where density differences do not cause layering of the medication, has allowed for the development of devices on Earth to create these microcapsules and devices that will aid in the drug delivery using this technology. Progress continues towards clinical studies in cancer patients one day in the future.

Ongoing research of gravitational unloading supported by dry immersion technology allows for a broad spectrum of possible clinical applications such as the early diagnosis of slow-developing neurological disorders, the combating of edema that responds poorly to medication, post-operative rehabilitation, sports medicine and rehabilitation for premature babies.

High-quality protein crystal growth experiment aboard Kibo

There are more than 100,000 proteins in the human body and as many as 10 billion in nature. Every structure is different, and each one of them holds important information related to our health and to the global environment. Each protein has a unique and complicated structure, which is closely related to its function. Therefore, revealing protein structure leads to an understanding of its function. However, it is difficult to analyze protein structures here on Earth, where gravity interferes with optimal growth. The perfect environment in which to study these structures is space; microgravity means there is no convection to disrupt the liquid solution, nor is there precipitation to cause heavier molecules to sink. Therefore, protein molecules form orderly, high-quality crystals that provide optimal structures for study. Many crystals of various proteins have been created in the unique environment of space.

The Japan Aerospace Exploration Agency (JAXA) has conducted nine sessions of protein crystallization experiments since 2003 in the Zvezda service module and has developed techniques to produce high-quality protein crystals in space. Based on these techniques, JAXA executed six sessions of experiments for the first series for the High-quality Protein Crystal Growth experiment (JAXA PCG) in the Japanese Kibo module on the space station from July 2009 to May 2013. JAXA is conducting another six sessions in total as the second series by periodic flight opportunities of six-month intervals. The first session of the second-series experiments started in March 2014.

Through collaboration with the Russian Federal Space Agency (Roscosmos), protein samples are launched to the space station aboard the Russian Progress or Soyuz spacecraft. Soon after the docking, the samples are brought into Kibo to be placed inside the Protein Crystallization Research Facility (PCRF) where they

> Microgravity allows for optimal growth of the unique and complicated crystal structures of proteins leading to the development of medical treatments.

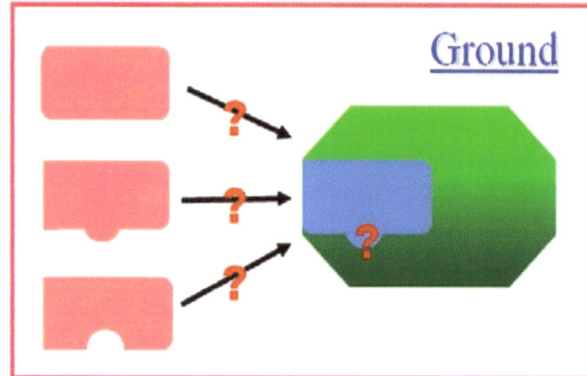
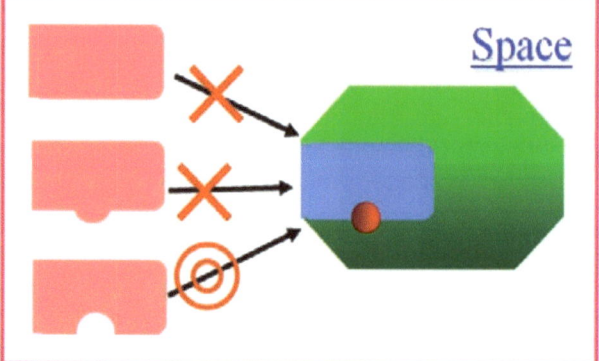

Advantage of the Space Experiment.

Because the structure of the disease-causing protein, or the keyhole, is vague when it is obtained on the ground, the shape of the key, or a medicine candidate compound for treatment cannot be determined. However, it is possible to find the structure of the disease-causing protein through the space experiments and medicine that fits the treatment (the key that fits the keyhole) can be developed.

Image credit: JAXA

are kept for a period of one-and-a-half to four months at a stable temperature, 68 degrees Fahrenheit (20 degrees Celsius). A counter-diffusion method called "Gel-Tube method" is used for crystallization whereby polyethylene glycol or salt solution is diffused into the protein solution separated by a porous membrane inside a tube. In this method, concentration of polyethylene glycol in the protein solution gradually increases and finally satisfies the condition for protein crystallization.

One of the major purposes of the protein crystal growth experiments is the contribution to the development of medical treatments. The relationship between a certain protein that causes disease and its medicine that suppresses the disease can be compared to the relationship between a "keyhole" and its "key." If the shape of the keyhole becomes apparent by examining the structure of the protein, treatment-oriented medicine with few side effects—the key to fit the keyhole—can be designed. JAXA is making positive advancements in research on obstinate diseases through experiments in space with the hope of supporting medical care more effectively.

An example of a protein that was successfully crystallized in space is hematopoietic prostaglandin D synthase (H-PGDS). This protein may hold the key to treating disease. A research team at the Osaka Bioscience Institute (OBI) reported that H-PGDS is expressed in certain muscle fibers of patients with Duchenne muscular dystrophy (DMD). An inherited muscle disorder, DMD is the most common form of muscular dystrophy, affecting approximately 1 in 3,500 boys. DMD causes muscular wasting and accelerates the progression of muscular deterioration. It is an obstinate disease for which a fundamental mode of treatment has not yet been found. Therefore, H-PGDS-specific inhibitors are considered to be useful drugs for muscular dystrophy.

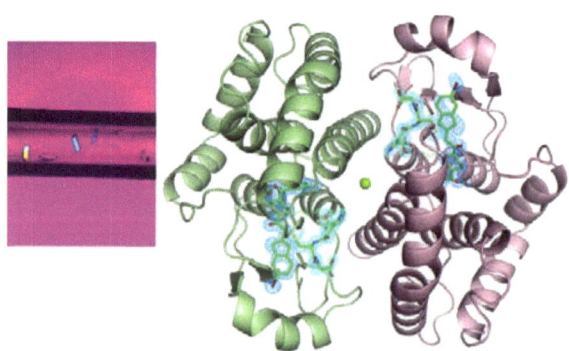

High-quality crystals of H-PGDS-Inhibitor complexes.

The detailed structure of muscular dystrophy related-protein became clear through a space experiment.

Image credit: Osaka Bioscience Institute/ Tsukuba University/Maruwa Foods and Biosciences, Inc./JAXA

The OBI research team has successfully determined the 3-D structure of H-PGDS in a complex with a prototype H PGDS-specific inhibitor. H-PGDS has been crystallized several times in microgravity as part of JAXA's space experiments. Using X-ray crystallographic analysis—using X-rays to determine the structure—researchers determined the structure of the high-quality crystals of H-PGDS-inhibitor complexes grown in space, and as a result, discovered a new inhibitor with several hundred times stronger activity than the prototype inhibitor. This particular experiment is an example of how understanding a protein's structure can lead to better drug designs. Further research is ongoing.

Cancer-targeted treatments from space station discoveries

Invasive and systemic cancer treatment is a necessary evil for many people with the devastating diagnosis. These patients endure therapies with ravaging side effects, including nausea, immune suppression, hair loss and even organ failure, in hopes of eradicating cancerous tissues in the body. If treatments targeted a patient's cancerous tissues, it could provide clinicians with an alternative to lessen the delivery of toxic levels of chemotherapy or radiation. Remarkably, research that began in space may soon result in such options here on Earth.

Using the distinctive, microgravity environment aboard the International Space Station, a particular series of research investigations is making further advancements in cancer therapy. A process investigated aboard the space station known as microencapsulation is able to produce tiny, liquid-filled, biodegradable micro-balloons containing specific combinations of concentrated anti-tumor drugs. Using specialized needles, doctors could deliver these micro-balloons, or microcapsules, directly to specific treatment sites within a cancer patient, effectively revolutionizing cancer treatment.

> The microgravity environment has allowed for the development of devices on Earth to create microcapsules that could aid in drug delivery.

Dr. Dennis Morrison of NASA's Johnson Space Center used the microgravity environment aboard the space station for microencapsulation experiments as a tool to develop the Earth-based technology, called the Microencapsulation Electrostatic Processing System-II (MEPS-II), to make the most effective microcapsules. The technique for making these microcapsules could not be done on Earth, because the different densities of the liquids would layer. But in space, microgravity brought together two liquids incapable of mixing on Earth (80 percent water and 20 percent oil) in such a way that spontaneously caused liquid-filled microcapsules to form as spherical, tiny, liquid-filled bubbles surrounded by a thin, semipermeable, outer membrane.

In space, surface tension shapes liquids into spheres. Each molecule on a liquid's surface is pulled with equal tension by its neighbors. The closely integrated molecules form into the smallest possible area, which is a sphere. In effect, the MEPS-II system allowed a combination of liquids in a bubble shape because surface tension forces took over and allowed the fluids to interface rather than sit atop one another. Studying the samples upon return to Earth allowed scientists to understand how to make a device that could create the same microcapsules on Earth.

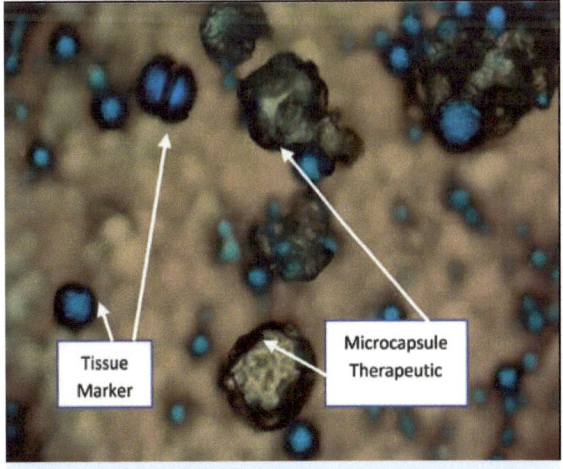

The oil contains a visualization marker that is traceable by ultrasound and CT scans to allow doctors to follow the microcapsules as they are site-specifically delivered to the tumor. The semipermeable outer skin releases the drug slowly, through its physical ability to be timed released.

Image credit: NuVue Therapeutics, Inc.

Dr. Dennis Morrison poses with the Microencapsulation Electrostatic Processing System flight hardware that was used on the International Space Station to produce microcapsules for cancer treatment delivery.

Image credit: NASA

The MEPS-II system is now being brought to commercial scale under U.S. Food and Drug Administration (FDA) Good Manufacturing Practice requirements, and commercialization of the MEPS technology and methods to develop new applications for these unique microcapsules has already begun. The space station research led to 13 licensed microcapsule-related patents and two that are pending.

In laboratory testing, MEPS-II microcapsules containing anticancer drugs were injected directly into a human prostate and lung tumors in animal models. These models were then, in follow-on tests, also injected following the delivery of specific cryo-surgical effects, similar to a freeze and thaw effect on the tumorous tissues. Injecting the microcapsules directly into the tumor demonstrated improved site-specific therapeutic results and the inhibition of tumor growth. Following cryo-surgery, the microcapsules demonstrated improved destruction of the tumor better than freezing or local chemotherapy alone.

Though Morrison's previous laboratory studies of microcapsules were primarily focused on prostate and lung cancer, his studies now target breast cancer for the FDA approval process through development with NuVue Therapeutics, Inc. Though it will take a few years to get approval to use the microcapsules as a treatment option filled with anti-tumor drug therapies, several devices that will aid in drug delivery using this technology are planned for pre-clinical study as early as 2015.

After achieving full FDA approval, planned clinical trials will involve injecting the microcapsules with the anti-tumor drugs directly into tumor sites in humans at both MD Anderson Cancer Center in Houston and the Mayo Cancer Center in Scottsdale, Ariz. Given the success in animal models in laboratory studies with human prostate and lung tumor treatment, Morrison has high hopes in the near future of being able to begin use of the microcapsule treatment in breast cancer.

These kinds of technologies are enabled by the availability of the microgravity environment aboard the space station. Just as microgravity can aid in the discovery of new technologies for cancer treatment, these microcapsules may one day aid in the recovery of breast and other specific deep-tissue cancers.

Using weightlessness to treat multiple ailments

The technology of dry immersion was developed as an Earth-based model to study the effect of microgravity factors on the human body. Using this model, the effectiveness of measures developed to prevent the negative impact of spaceflight factors on people has been and continues to be evaluated. The countermeasures currently used on the International Space Station were tested in experiments involving dry immersion.

> Ongoing research on dry immersion technology allows for a broad spectrum of possible clinical applications.

Experts at the Institute of Biomedical Problems (IBMP) developed the automated immersion system to create water hypodynamia (utility model patent #44505 "Immersion bath" and invention patent #2441713 "Polymer covering and device for dry immersion"). The concept of the technology involves submerging a person in an immersion bath filled with water. The immersion system is an ergonomically designed tub with a built-in elevation mechanism, filtration, and temperature control systems. The subject is kept separate from the water by a thin water-proof cloth with an area significantly exceeding the area of the water's surface. In this way, conditions closely simulating the lack of gravity are recreated. As a result, changes typical of acute gravitational unloading are reproduced in the body.

The potential use of the system for health purposes relates to the specific physiological changes in the body caused by gravitational unloading. In particular, acute disruptions occur in the mechanisms of sensory interaction. These disruptions counteract compensatory processes in the central nervous system resulting in discovery of latent neurological disruptions. Treatment with dry immersion is also accompanied by a number of physiological shifts such as the redistribution of fluids in the body, which have positive effects in certain cardiovascular conditions such as edema.

The drug-free method of dry immersion offers the user: relaxation of muscles; increase in immunity; elimination of edema; and, normalization of blood pressure, thus making it possible to use the immersion system for the early diagnosis of slow-developing neurological disorders and to combat massive edema that responds poorly to pharmacological remedies. Its use may also be an effective mechanism in rehabilitative treatment in areas such as psychoneurology, traumatology, orthopedics (post-operative rehabilitation), sports medicine, clinical neurophysiology, and applied psycho-physiology.

The use of the immersion system is also a particularly valuable rehabilitation measure for premature babies who are exposed to the effects of gravity following the intrauterine environment. Perinatal damage to the central nervous system (hyperexcitability; depressive, muscle hypertonia, and cephalohematoma syndromes) is an opportunity for the use of the dry immersion method. Additional uses for dry immersion include treatment of immune disorders, hormone imbalances, muscle disease, wound healing, and cardiovascular health.

The spectrum of possible applications of this system that simulates spaceflight conditions, such as those experienced by the International Space Station, is fairly broad and will expand with further study.

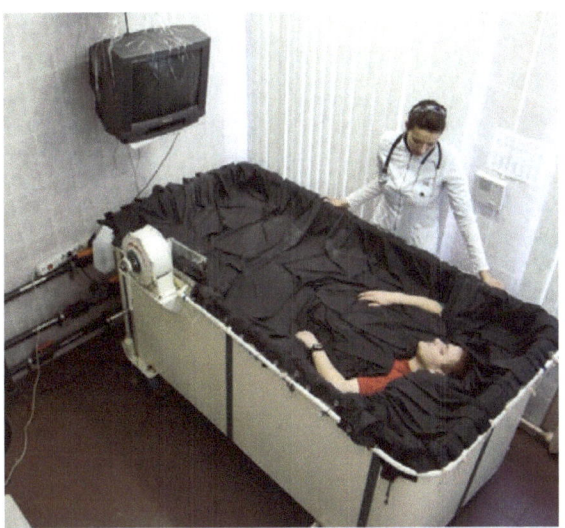

Dry Immersion Complex (prototype).
Image credit: Institute of Biomedical Problems, Russia

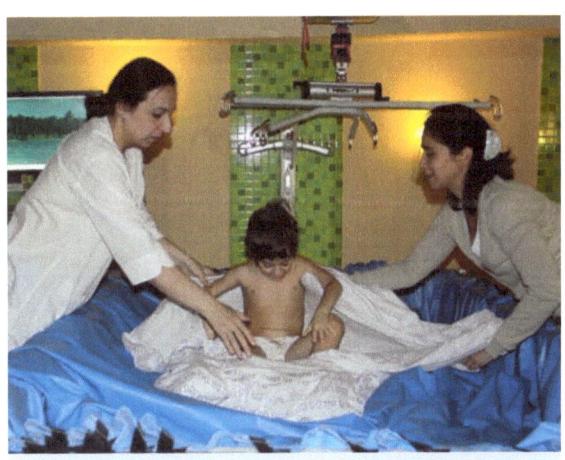

New prototype of the automated immersion system used in clinics.
Image credit: Aerospace Medical Center and Technology

Food and the Environment

Microbiology is a vitally important area, not only within human spaceflight but also for humans on Earth. Microorganisms such as bacteria, archea, fungi and algae have a detrimental or a beneficial impact on our daily lives. This research has far-reaching effects feeding into many different areas of biotechnology as microorganisms have a role in food spoilage, waste and sewage treatment and processing, nutrient cycling and exchange, pollution control, and in increased greenhouse gases.

Studying the effects of gravity on plants led to the development of an ethylene scrubber. This technology is now used as an air purifier that destroys airborne bacteria, mold, fungi, viruses, and odors. The scrubber is used for food preservation in major supermarkets, high-end refrigerator technology, and in trucks that carry groceries to remote regions of countries such as India, Saudi Arabia and Kuwait to name a few. Even the health care industry benefits from the use of these units in clinics, operating rooms, neonatal wards and waiting rooms making these locations safer for their inhabitants.

Plant research in a space greenhouse has allowed the study of root zone substrates in space allowing scientists to improve predictions of how artificial soils will behave when irrigated both in space and on Earth in experimental forests.

Microbiology applications from fungal research in space

Microbiology is a vitally important area, not only within human spaceflight but also for humans on Earth. With this microscopic world encompassing microorganisms such as bacteria, archea, fungi and algae that can have both a detrimental or beneficial impact on our daily lives, this research has far-reaching effects feeding into many different areas of biotechnology.

Microorganisms have both negative and beneficial effects, and different species of fungi are inherent in many of these processes.

- They can spoil food but assist in waste and sewage treatment and processing as well as nutrient cycling and exchange.
- Assist in pollution control but also increase greenhouse gases.

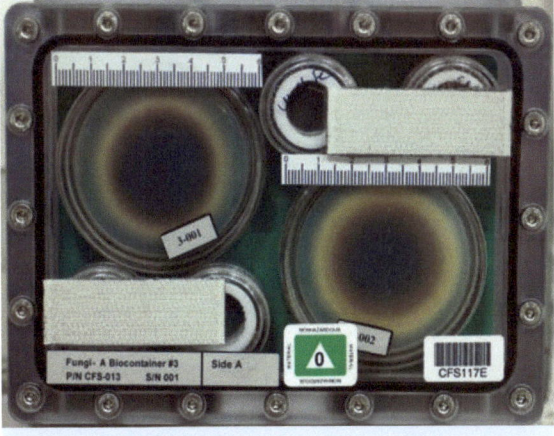

Flight day 5 sample from the Coloured Fungi in Space experiment.

Image credit: CFS-A science team

> Microbiology research on ISS provides insight into how microorganisms can spoil food, assist in waste and sewage treatment, and in pollution control.

- Cause disease but can be used in the manufacture of antibiotics, detergents and pesticides.
- Cause deterioration in manufactured materials and buildings but can also be used in the recover of metals in the mining sector as well as the production of biofuels and fertilizers.

Insights into one species may provide insights into others and hence feed into different applications.

The Growth and Survival of Coloured Fungi in Space-A (CFS-A) experiment determined the changes that weightlessness and cosmic radiation have on the growth and survival of various coloured fungi species. Understanding any changes in the physiology and survivability of different microorganisms in space can help determine the effect that this may have on spacecraft, associated systems and supplies, as well as the astronauts inhabiting them. This could provide important insights for developing countermeasures to possibly deleterious microorganisms, help to draw conclusions on how the space environment may affect similar organisms, and could also feed into biotechnology applications in the future.

The main fungal species studied in the CFS-A experiment was *Ulocladium chartarum*, which is well known to be involved in biodeterioration of organic and inorganic materials and suspected to be a possible contaminant in spacecraft. Other species studied were *Aspergillus niger* (which causes a disease called black mold on certain fruits and vegetables, and commercially accounts for 99 percent of global commercial citric acid production); *Cladosporium herbarum* (frequently the most prominent mold spore in air and found on dead herbaceous and woody plants, textiles, rubber, paper, and foodstuffs of all kinds); and *Basipetospora halophile* (which survives in high-saline environments).

The CFS-A experiment clearly indicated that *Ulocladium chartarum* is able to grow under spaceflight conditions, elaborating a new strategy to survive for a short time by developing submerged mycelium and for a long time by developing sporulating microcolonies on the surface of the nutrient source on which it was cultured. In spacecraft, *U. chartarum* and other fungal species could find a favorable environment to grow invasively unnoticed in the depth of surfaces under the right conditions posing a risk factor for biodegradation of structural components, as well as a direct threat for crew health. This will be especially important for future long-duration missions outside of low-Earth orbit where astronauts will have to be more self-sufficient for maintaining spacecraft and systems, and some food supplies would need to be preserved for longer than potentially 18 months. However, on the same line, in the future this kind of research could potentially feed into strategies for waste recycling on spacecraft and the development of biological life support systems. As we gain knowledge of the life histories of key species of fungi in the space environment, that knowledge can be readily applied to better manage these species on Earth.

Fungi on the ISS, grows on a panel of the Russian Zarya Module where exercise clothes were hung to dry.
Image credit: NASA

Plant growth on ISS has global impacts on Earth

Understanding the effects of gravity on plant life is essential in preparation for human exploration beyond low-Earth orbit. The ability to produce high-energy, low-mass food sources during spaceflight will enable the maintenance of crew health during long-duration missions while having a reduced impact on resources necessary for long-distance travel.

The Advanced Astroculture™ (ADVASC) investigation, led by Weijia Zhou, Ph.D., of the Wisconsin Center for Space Automation and Robotics, University of Wisconsin-Madison, explored the benefits of using microgravity to create custom crops that can withstand the inhospitable climates of space, resist pestilence, and need less volume to grow. ADVASC was performed over several International Space Station (ISS) expeditions, growing two generations of Arabidopsis thaliana (rapidly growing, flowering plant in the mustard family that has been grown on many space missions), and soybean plants, from seed to seed in space using the ADVASC payload, an autonomously operated plant growth unit. The ability to grow plants from seeds through several generations has proven to be challenging in space and is critical in developing hardware and operational concepts to take human explorers farther beyond low-Earth orbit.

> A new technology for an ethylene scrubber is used for food preservation in major supermarkets and in trucks that carry groceries to remote regions of the world.

While serving as a unique plant-growth chamber, the ADVASC hardware design has also contributed to national security, cancer-fighting pharmaceuticals and educational tools for students. ADVASC's novel air scrubber was designed to remove ethylene from the chamber atmosphere, thus allowing longevity of the produce. Ethylene is a naturally occurring, odorless, colorless gas given off by plants that hastens the

Astronaut Peggy Whitson with the ADVASC soybean plant growth experiment during Expedition 5.
Image credit: NASA

ripening of fruits and the aging of flowers, encouraging decay. Comprised of carbon and hydrogen in closed growing environments, like on a spacecraft or in a terrestrial greenhouse, ethylene builds up quickly and plants mature too fast. Removing ethylene, therefore, is important to preserving crops not just in space, but also on Earth, where grocers and florists have an interest in reducing the gas that ultimately shortens the shelf life of their products.

The ethylene-reduction device, also called the ethylene "scrubber," draws air through tubes coated in thin layers of titanium dioxide. The insides of the tubes are exposed to ultraviolet light, which creates a simple, chemical reaction, converting the ethylene into trace amounts of water and carbon dioxide, both of which are actually good for plants.

KES Science & Technology Inc., a Kennesaw, Georgia-based company specializing in sustaining perishable foods, licensed the ethylene-scrubbing technology from the University of Wisconsin. KES partnered with Akida Holdings, of Jacksonville, Florida, which now markets the NASA-developed technology as AiroCide. According to the company, it is the only air purifier that completely destroys airborne bacteria, mold, fungi, mycotoxins, viruses, volatile organic compounds (like ethylene), and odors. What's more, the device has no filters that need changing and produces no harmful byproducts, such as the ozone created by some filtration systems.

Food preservation customers include supermarkets like Whole Foods; produce distribution facilities like those operated by Del Monte; food processing plants; wineries; distilleries; restaurants; and large floral shops. Reeves Floral, an AiroCide user, reported 92-percent reductions in airborne mold and a 58-percent drop in airborne bacteria levels in just the first 24 hours it had the units operating in its floral storage warehouse. The AiroCide units can be used in walk-in coolers to preserve freshness of produce during storage and transport, to increase safety in food preparation areas, to kill bacterial contaminants in flowers (botrytis), and to protect against spoilage and contaminants.

AiroCide technology is now incorporated into a line of refrigerators, high-end consumer models that preserve freshness and reduce food waste. The refrigerator recycles the air every 20 minutes, reducing odors, viruses, and bacteria, as well as eliminating the presence of veggie-wilting ethylene.

AiroCide units have been deployed to India and the Gulf Cooperation Council, which includes the countries of Bahrain, Kuwait, Qatar, Oman, Saudi Arabia, and the United Arab Emirates. In these areas, where refrigerated trucks carry groceries from rural farmland to towns miles away, the AiroCide unit preserves freshness and prevents food spoilage in harsh environments.

In the health care arena, AiroCide units have been incorporated into doctors' clinics, operating rooms, neonatal wards, and in waiting areas, an often overlooked location rife with germs and bacteria like respiratory influenza or mycobacterium tuberculosis and frequented by people with compromised immune systems. Operating rooms with AiroCide units mounted in the ceiling become safer for all inhabitants, as harmful bacteria like methicillin-resistant *Staphylococcus aureus* and vancomycin-resistant *Enterococcus*, and the fungi *Penicillium* and *Aspergillus* are removed from the air. In addition to eliminating virtually all known airborne germs and diseases, the technology reduces the burden on high-efficiency particulate air filters and laminar flow environments. These same air-cleaning properties have also been applied to neonatal wards.

The AiroCide units have been adapted for use in everyday living environments. In hotels, for example, the units eliminate mold, mildew, germs and unwanted odors. These same features are also useful in offices, where illnesses caused by airborne organisms can lower productivity. In homes, the AiroCide units help eliminate the growth of mold and fungi as well as eliminate allergens like pet dander and dust mites.

Experiments with higher plants on the Russian Segment of the International Space Station

Some of the most important tasks in space biology include the creation of reliable and effectively functioning life support systems, and providing sustaining food sources for crew members. For long-term interplanetary spaceflights and planetary bases, the human life support system and food production

> Space station studies improved predictions of how artificial soils will behave when irrigated both in space and on Earth in experimental forests.

has to be based on regenerating the living environment from life support products through physical/chemical and biological processes. Greenhouses will most likely be designed for the cultivation of vegetables, primarily greens and herbs. However, in order to implement these plans, plants must grow, develop, and reproduce in spaceflight with cultivation productivity similar to Earth. To address this need, a series of 17 *Rasteniya* experiments were conducted from 2002-2011 using the Lada greenhouse on the Russian segment of the International Space Station.

Multigenerational studies were carried out to culture genetically tagged dwarf pea plants in the Lada space greenhouse. For the first time in space research, four consecutive generations of genetically tagged pea line seeds were obtained in spaceflight. The growth and development characteristics of various lines of pea plants did not change in a significant way compared to ground control samples. Using molecular methods with random amplified polymorphic DNA (RAPD) primers with 10 markers and analyzing chromosomal aberrations, it was demonstrated that plants having undergone four complete development cycles in spaceflight did not manifest genetic polymorphism. That makes it possible to assert that there is no impact of spaceflight factors on the genetic apparatus of plants in the first to the fourth "space" generations.

To prepare a chain of higher plants for future life support systems of space crews, experiments were carried out to cultivate the leafy vegetable plant mizuna *(Brassica rapa var. nipposinica)*. Results showed that the significant increase in the parameter of total contamination of International Space Station (ISS) air did not result in a decrease in productivity of the leafy vegetable plant; however, the plants responded with a change in gene expression.

A space experiment to grow super dwarf wheat during a complete vegetation cycle showed that the

Cosmonaut Valery G. Korzun, Expedition Five mission commander, studies mizuna lettuce as part of the Rastenyia-2 investigation.
Image credit: NASA

rate of plant development over 90 days did not differ from data from ground control experiments. When the space-produced seeds were planted on the ground, plants that grew were no different from the control sample.

The work done has great applied value because in the process of creating and operating the space greenhouse, cutting-edge equipment and software were developed, making it possible to grow plants automatically. This dual-purpose technology can also improve plant growth on Earth. The psycho-physiological aspect of the interaction between humans and plants in a habitable pressurized volume was studied, and data were obtained on the safety of cultivating plant biomass on a space station for human consumption. These data are of great interest for design work to create productive greenhouses that are part of promising life support systems of any living complexes that are cut off from the Earth's biosphere.

Scientists have also studied the interaction of plants with the soil. The processes by which plant roots receive water, gases and nutrients are different in space than they are on Earth. On Earth, gravity and surface tension combine to move water through soil, allowing air to move through the pore spaces in the soil to the plant's roots. In space, soil is replaced with an artificial growth medium, made up of small grains or other porous material. In microgravity, liquid moves through capillary action, where the liquid is attracted to the adjacent surface of a solid material. The surface tension of the liquid pulls additional liquid along as each new surface is wetted. If the plant is over-watered and all of the surface area and open spaces within the growth medium are filled with liquid, then gas (air) can't move, and the plant's roots are deprived of air and oxygen. When properly wetted, as water is used by the roots, surface tension pulls additional liquid along without filling the pore spaces, and therefore without preventing oxygen from diffusing through the open spaces to the roots. Studies in the Lada greenhouse have addressed the importance of root zone media in these extreme artificial conditions. Scientists have studied a variety of root zone substrates—growth media, material particle sizes, and packing structure—and learned which combinations work best.

Knowledge of root zone substrates in space has allowed scientists to improve their predictions of how artificial soils will behave when they're irrigated—in space and on Earth—and to design specific plant growth media and artificial soils for greenhouses and other large scale plant production facilities on Earth. Models, describing the behavior of water and oxygen learned from these space experiments, have been published in scientific journals, allowing commercial users to access the information without divulging their propriety growth media mixtures. Sensor technology developed to monitor the Lada root zone is being applied to monitor soil properties in a state-of-the-art measurement facility at an experimental forest.

Editor's note: *Colleagues from many Russian and non-Russian organizations participated in carrying out work according to the Rasteniya program in the Lada greenhouse on the ISS RS. The contributions of G. E. Bingham (Utah State University, Space Dynamics Laboratory, Logan, Utah, USA), S. A. Gostimsky (M. V. Lomonosov Moscow State University), and M. Sugimoto (Okayama University, Institute of Bioresources, Okayama, Japan) should be especially noted.*

Cosmonaut Maxim Suraev, Expedition 22 Flight Engineer, holds Mizun lettuce plants from the BIO-5 Rasteniya-2 ("Plants-2") experiment in the Service Module during Expedition 20.

Image credit: NASA

Heart Health and Biorhythms

Studying spaceflight effects on the cardiovascular system has led to the creation of unique instruments that can be used on Earth for the detection of the earliest deviations in health status. These technologies are now used to examine motor vehicle drivers and civil aviation pilots to evaluate risks and prevent accidents. Twenty-four-hour ECGs of astronauts were also analyzed to understand the space environment's effect on biological rhythm and cardiac autonomic nervous activity leading to recommendations for maintaining a well-balanced biological rhythm on Earth. One of these recommendations is maintenance of a regular sleep schedule. In studying the sleep patterns of cosmonauts using a miniature device that fits in their pocket, information is recorded and sent to Earth for analysis of sleep quality. An Earth model of this device is placed under the pillow or mattress to record movements related to heart and breathing.

Space cardiology for the benefit of health care

The cardiovascular system plays an exceptionally important role in cosmonauts' physical adaptation to long-term weightlessness. Since 2002, the scientific experiment "Puls" and from 2007 through 2012, the experiment "Pneumocard" were performed regularly on the International Space Station to study spaceflight effects on the cardiovascular system. These studies have provided a tremendous amount of information about space cardiology that has resulted in new technologies successfully used to evaluate the body's functional reserves, to determine the degree of stress on regulatory systems and to assess the risk of development of disease. These new technologies served as the basis both for further development of cardiological systems on the International Space Station and for the creation of unique sets of instruments that can be used in health care practice, including the Ecosan-2007 hardware-software complex.

> Instruments used on Earth for the early detection of deviations in health status are now used to examine motor vehicle drivers and civil aviation pilots.

Ecosan-2007 is a multi-purpose instrument for early detection of the earliest deviations in health status. It is based on the principle of prenosological diagnosis, which arose in space medicine, referring to the study of changes in the body that precede their development. This device is now used to examine drivers of motor vehicles, civil aviation pilots, and test subjects in experiments on Earth involving various stress factors.

A study that used the Ecosan-2007 to detect early health issues among 105 bus drivers showed that more than 30 percent of the drivers were in prenosological and premorbid states, which sharply increases the risk of motor vehicle accidents.

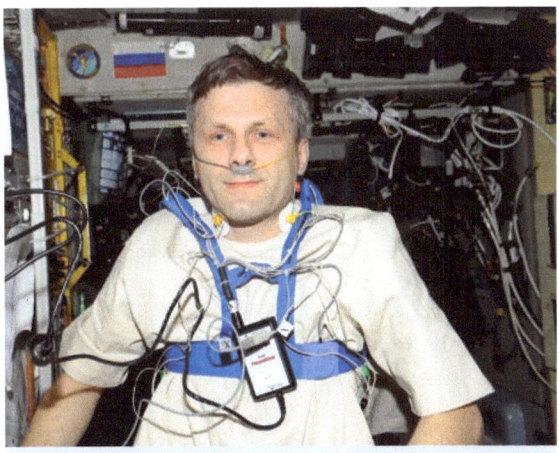

Andre Borisenko performs the Pneumocard experiment aboard the International Space Station.
Image credit: Roscosmos

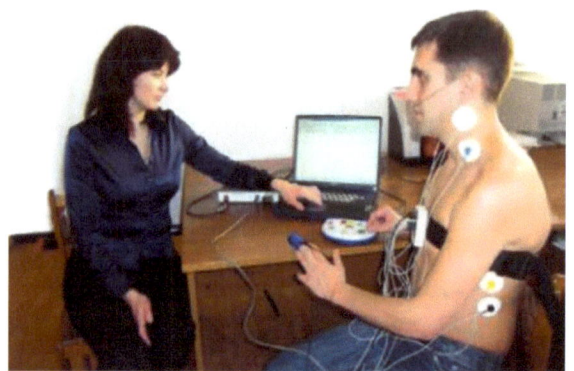

Research using the Ecosan-2007 complex.
Image credit: Institute of Biomedical Problems of the Russian Academy of Sciences

During examinations of civil aviation pilots the use of Ecosan-2007 showed that long-term, work-related chronic stress increases the risk development of pathologies, which should be considered during the expert evaluation of fitness for flight, especially for people at the age more than 50.

The Ecosan-2007 complex was also used in a 520-day experiment on Earth simulating a flight to Mars. During the experiment, monthly examinations of the "Martian" crew located in a pressurized mockup of an interplanetary spacecraft and at the same time of volunteer test subjects in control groups in 12 different regions of the world were performed. Long-term telemedicine of medical-environmental research using the Ecosan-2007 complex will be the prototype of a future system of individual prenosological monitoring, which will be based on space cardiology methods.

The results of the studies on Earth performed using the Ecosan-2007 complex served as a tool for Earth-based clinical use and as the basis for the further development of space cardiology technologies. Two new instruments have been developed for the space station and have both been in use since 2014. One of them, Cardiovektor, will be the advanced development of Pneumocard, which will make it possible to perform precise measurements of energy of heart muscle and to evaluate the activity of the right and left chambers of the heart. The second instrument, Cosmocard, will develop the methodology that was used in the Ecosan-2007 for electrocardiogram dispersion mapping. This will allow us to use the non-invasive study of the energy-metabolic characteristics of the cardiac muscle at various stages of spaceflight. In the future, these instruments may also be successfully used in health care practice.

Biological rhythms in space and on Earth

Humans wake up in the morning and fall asleep at night. The biological clock generates this regular, daily rhythm. Similar to the clock in the brain, there is a peripheral clock in every single cell in our bodies. The clock in the brain is called the "master clock," and the clock in cells is referred to as the "peripheral clock." The "master clock" and "peripheral clocks" communicate with one another through the autonomic, especially the sympathetic, nervous system to accurately regulate the biological rhythm.

It is not uncommon to perform docking and destination landings in the late night or early morning for maximizing efficiency during rocket launching or returning to Earth. In such cases, the astronauts experience an effect much like traveling through time zones during international travel. Furthermore, we also have broad biological spectra (i.e., 90-min, 8-hour and 7-day, etc.) and have developed "chronomics" to analyze them. Chronomics is the term that defines a new scientific field such as genomics and proteomics and can reveal hidden signals in the original time

> ISS research has led to recommendations for maintaining a well-balanced, biological rhythm on Earth.

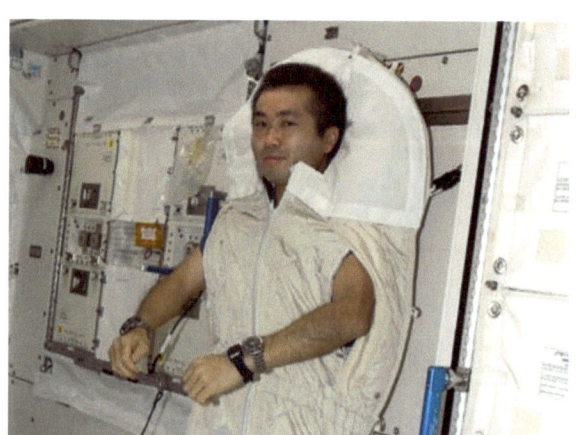

Space sleep – Crew members usually sleep in sleeping bags.
Image credit: JAXA/NASA

Sunrise from the International Space Station. Since the space station orbits the Earth once every 90 minutes, it sees a sunrise or a sunset every 45 minutes.
Image credit: NASA

series data, which are not decipherable without use of chronomics. Chronomes are time structures consisting of (1) multi-frequency rhythms, (2) elements of chaos, (3) trends in chaotic and rhythmic endpoints, and (4) noise (or as yet unresolved variables). Chronomics can solve the complexity of organism's interactions with their environment, near and far. The International Space Station (ISS) circles around the Earth once every 90 minutes, and astronauts aboard see sunrise and sunset every 45 minutes. The effects of high-intensity light deficiency on human health in the unique environment created by microgravity and 45-minute day and night cycles is not well understood.

There may be an increased risk for biological rhythm deregulation and insomnia when living in a space environment with microgravity and is deficient in high intensity light. We analyzed 24-hour ECG recordings of long-term-spaceflight astronauts with the aim to clarify the space environment's effect on biological rhythm and cardiac autonomic nervous activity. A small, lightweight Holter monitor was used to record 24-hour ECGs, preflight (once), inflight (three times), and postflight (once). The 24-hour ECG data recorded in orbit was downlinked from Kibo Japanese Experiment Module to Tsukuba Space Center for time domain and frequency domain analysis. As a result, we found that although biological rhythms of the astronauts were immediately disturbed after arrival on the ISS, they recovered in five months after the beginning of long-term stay, and their biological rhythms became well-regulated when compared to the preflight and postflight recordings. We speculated that although the astronauts experienced a time-difference effect during their long-term spaceflight after spacecraft docking, they were able to maintain a well-regulated biological rhythm because of their regular, daily schedule aboard the ISS.

We are exposed to environments that easily disturb our biological rhythm. Progress in communications, i.e., internet and smartphones, and transportation, i.e., airplanes, have made borders of countries and time zones irrelevant. It has been reported that disturbance in biological rhythm may cause diseases such as high blood pressure, obesity, hypercholesterolemia, diabetes, osteoporosis, insomnia, depression, premature aging, cancer, etc.

Three factors for maintaining a well-balanced biological rhythm are morning light, melatonin secreted at night, and a balanced diet (particularly breakfast). Most important is the exposure to bright light during daytime and complete darkness during sleep at night. The astronauts' daily regimen in maintaining a well-balanced biological rhythm during their long-term spaceflight serves as a guideline for us living on Earth. In particular, we recommend the following six activities for maintaining a well-balanced biological rhythm: 1) keep a regular daily schedule; 2) keep regular sleep and meal schedules; 3) drink coffee or green tea in the morning; 4) maintain appropriate temperatures during day and night; 5) engage in activities that support the peripheral clock such as exercise and getting dressed after waking; and 6) occasionally note the time.

Innovative space-based device promotes restful sleep on Earth

There has been a growing interest in the study of night-time sleep in weightlessness since the first steps of space exploration. Indeed, normal, good-quality sleep is the basis for maintaining the necessary, high psychological functioning and good physical condition of cosmonauts. However, the clinical method used to study sleep (polysomnography) requires the use of a large number of sensors, which disturbs normal sleep, requires time, and is too complex for spaceflight.

> A miniature device can be placed under a pillow or a mattress to record movements related to the heart and to breathing.

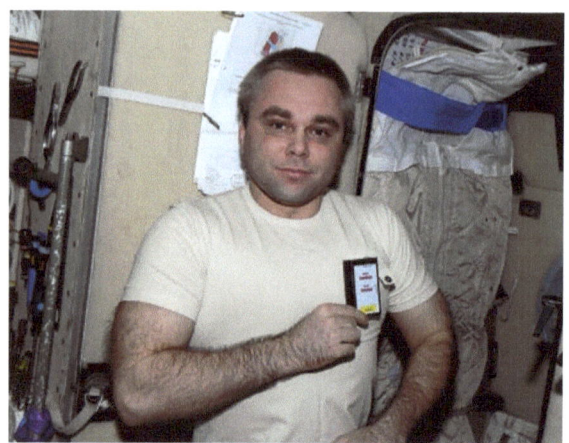

Crew member sets up the Sonocard device before sleep in spaceflight.

Image credit: Roscosmos

Thus, nowadays the data on the sleep quality of cosmonauts in long-term weightlessness have been extremely limited.

In 2007, a new device, Sonocard, arrived on the Russian segment of the International Space Station (ISS), making it possible to study sleep with a miniature device in the size of a deck of cards. The device is placed on the left in the top pocket of the cosmonaut's T-shirt before sleep, and its sensor elicits microfluctuations of the chest wall that are related to heart function. Upon wake-up, the information recorded during the night is sent to Earth for analysis.

Sonocard provides a contactless recording of physiological signals, and its use does not require attaching electrodes or special sensors to the body. Instead, it acquires its data by recording all the vibrations that are elicited by the sensor/accelerometer. Pulse rate, breathing rate, movement activity, and heart rhythm variability are obtained. This method is successfully used in various fields of medicine and physiology to assess the state of the basic body functions. Space medicine was one of its first fields of application, and to date, a large amount of experience has been accumulated on its use to assess the functional condition of cosmonauts during spaceflight. When analyzing the data obtained during sleep, the changes in the activity of the sympathetic and parasympathetic regulation chains in first hours after falling asleep and last hours before waking up are determined. This makes it possible to assess to what extent the body was able to rest during sleep and how much it replenished the functional reserves that were spent the day before.

Scientific experiments using the Sonocard device are conducted on the space station on all Russian cosmonauts every two weeks. Over five years, a large amount of information on sleep in weightlessness has been gathered. For the first time, it is possible to discuss results that are not impacted by factors of workload and psycho-emotional stress that are always present during the day while carrying out science experiments under the normal flight program. A flight index sleep quality showed that the quality of sleep on average for the entire flight for all 22 participants in the experiment is 77.4 percent.

The Sonocard contactless method of sleep study that was created for use on the space station has been successfully used on Earth. The Earth model used a sensor that was designed in the form of a plate to be placed under the pillow or mattress to record a person's body movements related to heart and breathing rate. The signals recorded during the night are downloaded to a computer and analyzed according to the methods already proven in space research.

The new hardware/software system called Cardioson-3 was tested in a series of experiments on Earth, including a long-term, 520-day experiment simulating a flight to Mars. Unique research experience of the cosmonauts' functional state during sleep can have further development in two directions: creation of new, more effective systems of evaluating sleep in space for the simultaneous medical control of all crew members and development of similar devices for the control of the quality of sleep in the interests of public health care practice.

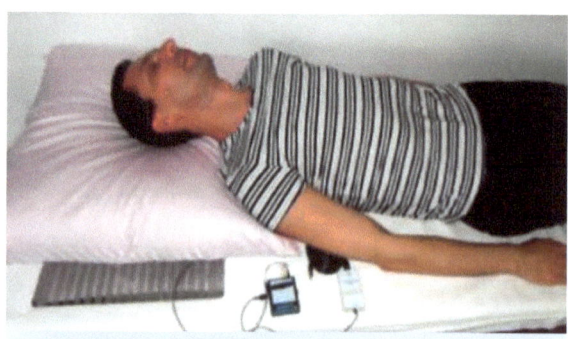

Cardioson-3 system–Earth analog of the Sonocard space device.

Image credit: Institute of Biomedical Problems of the Russian Academy of Sciences

Improving Balance and Movement

A new technology developed to correct motor disturbances in weightlessness has been used to treat patients with cerebral palsy, stroke, spinal cord injuries, balance problems and motor decline due to aging. Assessment of eye movement reactions of cosmonauts preflight and postflight has led to faster and less expensive diagnoses and treatment of patients suffering from vertigo, dizziness and equilibrium disturbances. A patented computerized, non-pharmacological method of preventing and correcting unfavorable perception and sensorimotor reactions is used to train patients and astronauts to acquire the ability to suppress vertigo, dizziness and equilibrium disturbances.

A system of hardware and software that collects information on body movements of astronauts on ISS has led to motor imagery protocols used in the research environment of a hospital in Rome in treatment of adult stroke patients and children with cerebral palsy. Other body movement research on ISS lead to the development of a suit for astronauts to compensate for the lack of daily loading from gravity. The clinical version of this suit is used for the comprehensive and drug-free treatment of cerebral palsy in children in Russia. Another clinical variation of this suit is used on patients who have suffered from stroke or brain trauma.

New technology simulates microgravity and improves balance on Earth

Spaceflight opportunities, such as that of the International Space Station (ISS), facilitated the development of Earth-based models of weightlessness and opened the door to studying the effects of the elimination of gravity. Over the 10 years since the ISS came into existence, a large amount of data, information and facts have been compiled by the Russian Institute of Biomedical Problems (IBMP) that have made it possible to switch from describing occurrences and phenomena to developing theories about the role and place of gravitational mechanisms in various bodily systems.

One example of an evolution in the development of new knowledge is the discovery of the leading triggering role in maintaining vertical posture, of

> A device to correct motor disturbances treats patients with cerebral palsy, stroke, spinal cord injuries, balance problems and motor decline due to aging.

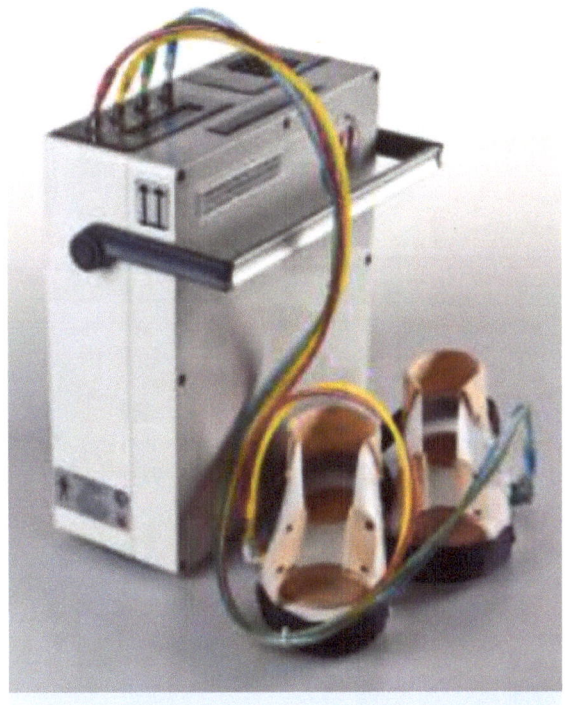

The support unloading compensator.
Image credit: Institute of Biomedical Problems

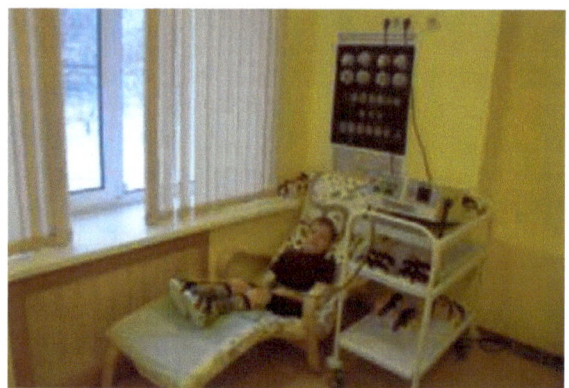

The support unloading compensator used on children.

Image credit: Aerospace Medical Center and Technology

sensory organs called Vater-Paccini corpuscles, located in the soles of the feet. These receptors were discovered back in the 19th century, but their role in gravireception was established very recently, thanks to ground-based simulation studies designed to keep cosmonauts' balance healthy in long-duration spaceflights. As a result of these studies, a unique piece of technology was developed to help to correct the motor disturbances in individuals with central nervous system dysfunction, injuries, balance problems, and motor decline due to aging.

Ground-based studies at IBMP identified a particular pattern of stabilization between body loading on the soles of the feet and subsequent motor stabilization processes associated with brain and spinal cord activities. The knowledge obtained revealed that the development of sensory-motor disturbances may be prevented by means of "artificial" support stimuli applied to the bottoms of the feet. The research results led to the development of new technology, the "support unloading compensator," a device that stimulates the support zones of the foot in natural human gait.

The uniqueness of this device lies in its ability to simulate the physical parameters that the support receptor or the foot receives during walking: the magnitude of pressure, temporal characteristics - duration of impact, intervals between stimulation of the heel and metatarsal support zones, and intervals between stimulation of the right and left foot.

But the disturbances of muscle tone and coordination of movement are not inherent exclusively to weightlessness. Diseases of the central nervous system (CNS),

brain injuries of various types, long-term motor unloading in aging, as well as some specific occupational conditions are also accompanied by the development of the aforementioned changes. Based on these data, IBMP scientists are working jointly with commercial companies to introduce motor disturbances corrective measures that were developed for weightlessness into the practice of treating and rehabilitating patients with profound motor lesions that are due to cerebral palsy, stroke, and brain and spinal cord injuries.

Between 2005 and 2011, the Center for Aerospace Medicine developed a clinical version of the support unloading compensator, the Korvit, and obtained all approval documents and licenses.

Today this technology is being successfully used in the most acute phase of stroke and facilitates more significant regression of motor disturbances and earlier recovery of locomotion than traditional treatment methods. Of particular interest are the data on prevention of muscle spasticity development of limbs affected by paresis in cases when the Korvit device is used in the first hours after the onset of a stroke.

The use of the support stimulation method in children during the early recovery phase after surgical treatment for fractures of the calf bones has facilitated the reduction of edema by 17 to 20 percent, and an increase in the range of motion in the ankle joint by 45 percent as early as the first 72 hours after surgery. Children who do not receive such treatment experience edema lasting six to eight days, which hinders motion in the injured limb and retards regenerative processes.

The use of the Korvit apparatus in the integrated rehabilitation of Cerebral Palsey patients has made it possible to maximize restoration of the balance of strength between extensor and flexor muscles, particularly in an upright position, and to normalize the functions of standing and walking, as well as the control of the coordination of various classes of movements.

New ways to assess neurovestibular system health in space also benefits those on Earth

Among the many problems that have confronted the medical sciences since humans first began exploring space, a main one is adaptation to the conditions of changed gravitational force. Upon arrival in weightlessness (first three to seven days) and upon the return to Earth (from landing to three to five days later), virtually all crew members experience a number of negative reactions and sensory disorders (orientational illusions, vertigo, dizziness, problems focusing on and tracking

> Assessment of eye movement reactions leads to faster and less expensive treatment of patients suffering from vertigo, dizziness and equilibrium disturbances.

visual objects), which are perceived as uncomfortable and can be accompanied by space motion sickness.

In weightlessness, information received from the vestibular apparatus within the inner ear does not align with information received from other sensory systems, so the typical sensory links are broken and the brain cannot correctly interpret the incoming signals at the beginning of flight, leading to the development of space motion sickness. As a result, this causes a decrease in the quality of crew member performance of work tasks, particularly those relating to visual tracking accuracy. It is very likely that the unsuccessful docking of spacecraft, errors in structural assembly, and other instances of errors in manual control that have occurred in orbit were often caused by disturbances in the function of tracking moving space objects because of changes in sensory functions.

Analysis of data accumulated in a series of scientific experiments before, during and after spaceflights on the stations Salyut-6, Salyut-7, and Mir has led the Institute of Biomedical Problems in Russia to develop a method that uses computerized systems named OculoStim-CM, Virtual and Sensomotor that can

Space experiment Virtual aboard the International Space Station (2013). Image credit: Roscosmos

Application of the Okulostim-KM hardware/software system in clinical studies.
Image credit: Institute of Biomedical Problems

accurately assess the state of vestibular function, intersensory interactions, spatial orientation, and visual tracking (Russian Federation patent #2307575 dated 10/10/2007, Kornilova L. N. et al.).

At the basis of the method lies the assessment of eye movement reactions – visual tracking tests that are conducted both with visual targets (stimuli) on a clean (black) field on the screen and against a backdrop of additional visual interferences (diffuse spots/ellipses moving horizontally or vertically) to "irritate" the peripheral vision. During the testing under differing conditions, movements of the eyes (by electro- and video-oculography) and head (using angular rate sensors and accelerometers) are recorded.

The eye-movement system is controlled by a complex hierarchy of innervation mechanisms located at different levels of the nervous system. The use of a special test battery makes it possible to evaluate the disruptions occurring in various forms of eye movements and, given the known mechanisms of how these movements are performed, to find the causes of these disturbances.

This method has been actively used during the preflight and postflight clinical/physiological examination of International Space Station (ISS) crew members in experiments called Virtual and Slezheniye (Pursuit) since 2013 and 2015, respectively and after crew members return to Earth during experiments Sensory Adaptation and Gaze-Spin since 2001 and 2009, respectively.

The procedure and hardware/software systems developed for spaceflight has also demonstrated effectiveness of the computerized method and hardware/software systems for use in diagnosing conditions of the vestibular and its related sensory systems (primarily visual) and in assessing the stability of static and dynamic spatial orientation on Earth. The systems have particularly been useful in experiments simulating weightlessness (immersion and bed rest); in examining highly qualified athletes (high-performance sports – gymnastics, figure and speed skating, target shooting, etc.); in diagnosing and treating patients suffering from dizziness and equilibrium disturbances; and, in evaluating the effectiveness of medications (betahistine drugs Betaver and Betaserc). The OculoStim-CM system was successfully certified for use in clinical studies of 200 patients with vertigo, dizziness and equilibrium disturbances, together with specialists from the nervous disease department of the I. M. Sechenov First Moscow State Medical University, the Academician Alexander Vein Clinic for the Treatment of Headaches and Vegetative Disorders, and the Federal Scientific Clinical Center of Otorhinolaryngology. Such application in clinical practice has made it possible to develop diagnostic criteria to determine the type of vestibular disturbance, while offering a rapid, less expensive initial differential diagnosis of dizziness and balance disturbances compared to traditional clinical testing.

Space research leads to non-pharmacological treatment and prevention of vertigo, dizziness and equilibrium disturbances

The history of spaceflight has shown that initial introduction to the weightlessness environment, such as that of the International Space Station, can lead to space motion sickness, making crew members feel dizzy and uncoordinated, and even impacting their ability to track objects with their eyes. The result can be a negative effect both on the health of crew members and the quality of their work performance in flight.

> Space research has led to a patented method used to train patients to acquire the ability to suppress vertigo, dizziness and equilibrium disturbances.

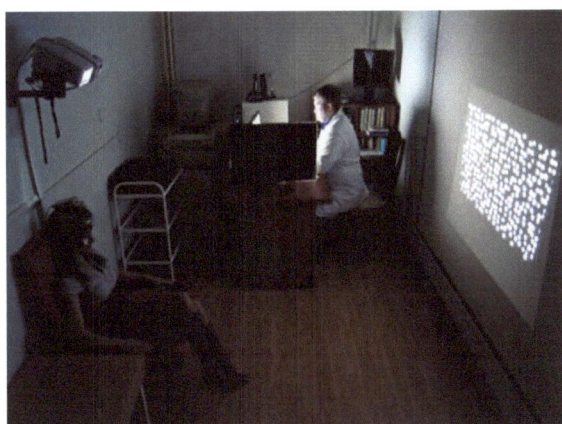

Visual method of training. (Laboratory of Vestibular Physiology, 2013)
Image credit: Institute of Biomedical Problems

At present, medications are typically used to eliminate the symptoms of space motion sickness, but they have a number of contraindications and side effects that can have a negative effect on various types of professional activity. Therefore, there has been an obvious need to develop non-pharmacological methods of preventing and treating space motion sickness.

It is well known that people in extreme professions, such as mountain climbers, athletes, acrobats, and ballet dancers develop the capability to suppress unfavorable vestibular reactions at the moment high accelerations act on them by developing a "fixation reflex." However, attempts by many clinicians to treat patients with vestibular problems using the same methods have been unsuccessful because of the vestibular challenges unique to this population. Therefore, experts at the vestibular physiology laboratory at the Institute of Biomedical Problems developed and patented a "Computerized method of preventing and correcting unfavorable perception and sensorimotor reactions" (Russian Federation patent #2301622 dated 06/27/2007, Kornilova L. N. et al.). The innovation of this method is in creating a unique approach to the training of patients depending on their disease (type of vestibulopathy) and in selecting the most effective means of training (visual, vestibular, or combined) for them using biofeedback.

During patient training, depending on the nature of the vertigo, dizziness or equilibrium disorder and of his/her disease (type of vestibulopathy), a series of training sessions is conducted to develop a unique fixation reflex using biofeedback, provided by the computer using this method to record eye and head movements.

Training is conducted until the negative reactions (vertigo, dizziness and equilibrium disturbances) the patient suffers from disappear or are significantly reduced. The therapeutic effect of the training is assessed through a follow-up clinical/neurological examination, including the use of the computerized method of comprehensively assessing the condition of vestibular function and visual tracking function (Russian Federation patent #2307575 dated 10/10/07, Kornilova L. N. et al.). The indicator of training success is the suppression of experimentally induced negative reactions (full or partial) during the action of visual and vestibular stimuli while fixing the gaze on an imagined target.

The non-pharmacological computerized method for treating and preventing vertigo, dizziness and equilibrium disturbances was tested in clinical conditions jointly with specialists from the nervous disease department of the I. M. Sechenov First Moscow State Medical University, the Academician Alexander Vein Clinic for the Treatment of Headaches and Vegetative Disorders, and the Federal Scientific Clinical Center of Otorhinolaryngology.

The results of the clinical work demonstrated that patients acquired the capacity to fixate on and hold the gaze on both real and imagined targets, thus suppressing (fully or partially) vertigo, dizziness, nystagmus, and equilibrium disturbances. It was shown that training effectiveness depended not only on the disorder (type of vestibulopathy), but also on the type of training selected. For patients with peripheral vestibulopathies, the most effective was visual training; for patients with central vestibulopathies, the vestibular method was best; and for patients with psychogenic vestibulopathies, the combined method was preferred.

Analysis of special questionnaires demonstrated that all patients with psychogenic, 91 percent of patients with peripheral, and 80 percent of patients with central vestibulopathies subjectively noted "good suppression of vertigo in everyday conditions" and "improvement in general adaptation to real life conditions."

The effectiveness of the non-pharmacological computerized method has made it a good candidate for use both during the preflight training of ISS crew members, and during spaceflight, to suppress the symptoms of space motion sickness. The suppression of negative reactions during flight using the fixation reflex has been successfully applied by crew members aboard the ISS since 2013.

Capturing the secrets of weightless movements for Earth applications

Leaving Earth's gravity initially impairs sensorimotor coordination in astronauts. Understanding how weightless astronauts learn to move and interact with objects will improve mission safety. There are also benefits to people on Earth because certain populations with brain injury or disease experience many of the same coordination challenges that astronauts experience.

The ELaboratore Immagini TElevisive - Space 2 (ELITE-S2) developed by Kayser Italia for the Italian Space Agency (ASI), is a system of hardware and software that collects information on body movements of astronauts on the International Space Station (ISS) with great accuracy. Two studies have been carried out with ELITE S2, Movement in Orbital Vehicle Experiments (MOVE) and Imagery of object Motion Affected by Gravity In Null-gravity Experiments (IMAGINE). MOVE involves reaching and touching a stationary target placed beyond arm's length while standing. This is a typical movement that on Earth would perturb balance unless it was compensated by adjusting the posture of trunk and leg muscles. While most healthy individuals on Earth can make sure movements, some patients, such as those with Parkinson's Disease, cannot. Interestingly, early in flight, astronauts make postural adjustments that are unnecessary in space but are part of their typical movement patterns on Earth. As the flight progresses, they learn to reduce unnecessary muscle activation, although with significant inter-subject variability.

IMAGINE tested the ability to interact with a moving object. People have implicit knowledge of how the world works since infancy, which allows them to react appropriately to the environment. In particular, gravity effects on an object in motion are detected accurately so that we can catch a falling object instantly. However, patients with a brain lesion, such as that which is due to stroke, often lack this ability. A previous study using an antecedent of ELITE S2 demonstrated that over the

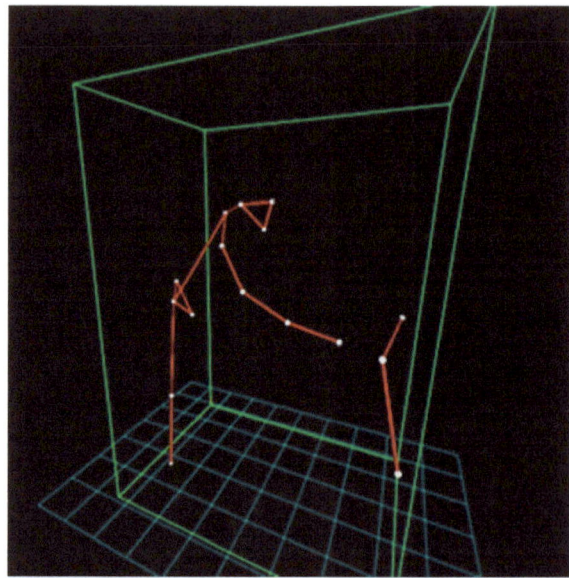

Limb and body movements recorded by ELITE-S2 during one experiment on the International Space Station.
Image credit: ASI

course of a two-week flight of the space shuttle, the motor responses of the astronauts were still sharply tuned to gravitational acceleration (http://science1.nasa.gov/science-news/science-at-nasa/2002/18mar_playingcatch/). In this study, astronauts asked to catch a ball that was projected "downward" from the ceiling contracted their limb muscles too early, anticipating non-existent gravity effects. Later, in the IMAGINE study, the astronauts were instead asked to throw an imaginary ball toward the ceiling and to catch it upon return in their hand. The motion of the ball was purely imaginary, but subjects performed an actual movement of their arm as realistic as possible to throw the ball. In different blocks of trials, they were asked to imagine either that the ball motion was affected by Earth's gravity (1 g) or that it was unaffected by gravity (0 g). Moreover, they were asked to vary the momentum of the throw from trial to trial. Surprisingly, it has been found that astronauts (including new flyers) are able to implicitly reproduce the 0 g conditions already on ground, prior to their mission, although their performance improved over the course of their time in space. There is a striking dissociation, therefore, between the motor control system that drives automatic responses tuned to Earth's gravity, where the anticipation of gravity persists even after two weeks in space, and

> **Motor imagery protocols used in the research environment of a hospital in Rome treat adult stroke patients and children with cerebral palsy.**

the cognitive system that allows one to imagine 0-g movements even on the ground and that appears endowed with a more general implicit knowledge of Newton's laws.

The results from this study will not only provide more information that can be used to keep astronauts healthy but could also lead to new rehabilitation strategies to help people with brain injuries. In fact, mental imagery represents a powerful tool to rehabilitate sensorimotor coordination in disabled patients. These motor imagery protocols are currently used in the research environment of the Neuromotor Rehabilitation Hospital of Santa Lucia Foundation in Rome. In particular, imagery training is used in adult stroke patients as well as in children with cerebral palsy. The training protocols involve the presentation of computer displays under supervision by medical or paramedical staff. Mental imagery is typically trained in combination with conventional physical therapy over a period of several weeks. Improved motor performance can be attributed to two main factors: 1) mental practice helps keep the motor programs active when little or no movements are possible, and 2) it allows an increase in the duration of the training session without adding to the physical demands of the task. There are still open issues in this ongoing research, such as establishing the best time window in the course of the disease at which mental practice could prove effective as well as determining the long-term effects of imagery training.

Space technologies in the rehabilitation of movement disorders

More than 50 years have passed since the first human spaceflight. As the duration of the flights has increased considerably, and amount of in-orbit activities has become greater, the need to maintain healthy bones and muscles in space has become more critical. Bones and muscles rely on performing daily activities in the presence of Earth's gravity to stay healthy. In space, traditional Earth-based methods to maintain bones and muscles, such as physical exercise, are challenging due to constraints that include such factors as crew time and vehicle size. To meet these challenges, specialists from the Institute of Biomedical Problems in Russia and their commercial partner, Zvezda, developed the Penguin suit to provide loading along the length of the body (axial loading) in a way that compensates for the lack of daily loading that the body usually experiences under the Earth's gravity. The first testing of the suit in space was performed in 1971 aboard the Salyut-1 station. Now the Penguin suit is actively used on the International Space Station as a regular component of the Russian countermeasure system of health maintenance.

> Treatment suits are used in Russia for the comprehensive and drug-free treatment of cerebral palsy in children and in patients with stroke and brain trauma.

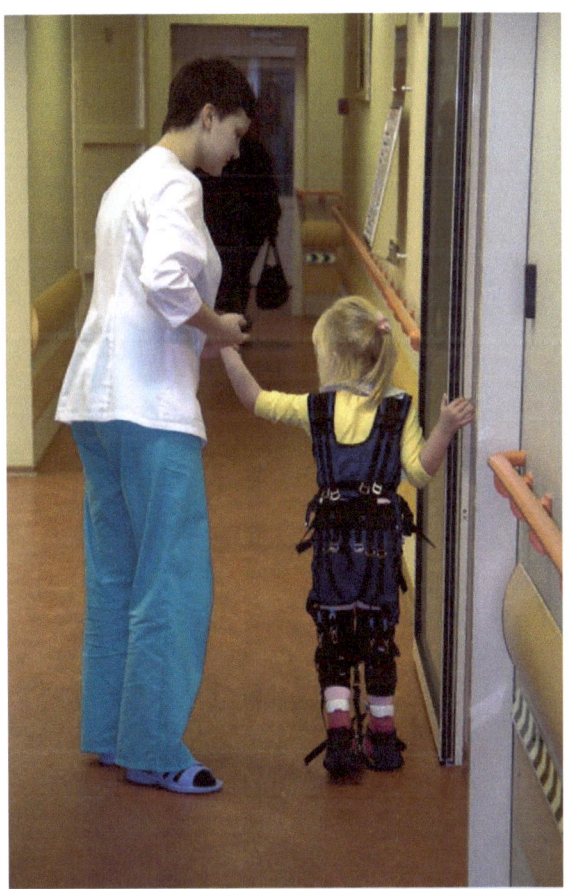

The Adeli treatment suit in use for pediatric rehabilitation.

Image credit: Aerospace Medical Center and Technology, Russia

Since the early 1990s, Professor Inessa Kozlovskaya and her team at the Institute of Biomedical Problems in Russia have implemented the use of this axial loading suit in clinical rehabilitation practice. The clinical version of the Penguin suit, the Adeli, was developed at the Institute of Pediatrics Russian Academy of Science under the leadership of Professor Ksenia Semyonova and is used for the comprehensive treatment of cerebral palsy in children. The treatment method is focused on restoring functional links of the body through a corrective flow of sensory information to the muscles, thereby improving the health of the tissues being loaded. This results in the correction of walking patterns and stabilization of balance in a relatively short period of time, including for those cerebral palsy children with deep motor disturbances. The Adeli suit was licensed in 1992 and has been continuously developed since. These methods have become one of the most popular and widely used in Russian medical clinics for rehabilitation of children with infantile cerebral paralysis.

New methods were also developed for patients undergoing motor rehabilitation after stroke and brain trauma. Paralytic and paretic alterations of motor functions that are the most frequent after-effects of these diseases typically lead to significant limitations in motor and social activity of these patients, decrease their functional abilities and obstruct their rehabilitation. Given all of the complexities and importance of the rehabilitation of these patients, another clinical modification of the Penguin suit was developed called the "Regent suit." The complex effect of the Regent suit on the body is based on an increase of the axial loading on skeletal structures and an increase in resistive loads on muscles during movement, which results in an increase of sensory information to the nervous system that is important for counteracting the development of pathological posture and for normalization of vertical stance and walking control. The Regent suit is effectively used at the early stage of rehabilitation for patients having movement disorders after cerebrovascular accident and cranium-brain traumas.

The clinical studies of the efficacy of the Regent suit in the rehabilitation of motor disorders in patients with limited lesions of the central nervous system were performed in acute and chronic studies with the participation of hundreds of stroke and brain trauma patients in the hospital № 83 Federal Medical-Biological Agency of Russia under leadership of professor Sergey Shvarkov, and in the Center of Speech Pathology and Neurorehabilitation under leadership of professor Vicktor Shklovsky. The efficacy of the suit in patients with post stroke hemiparesis was assessed at the Scientific

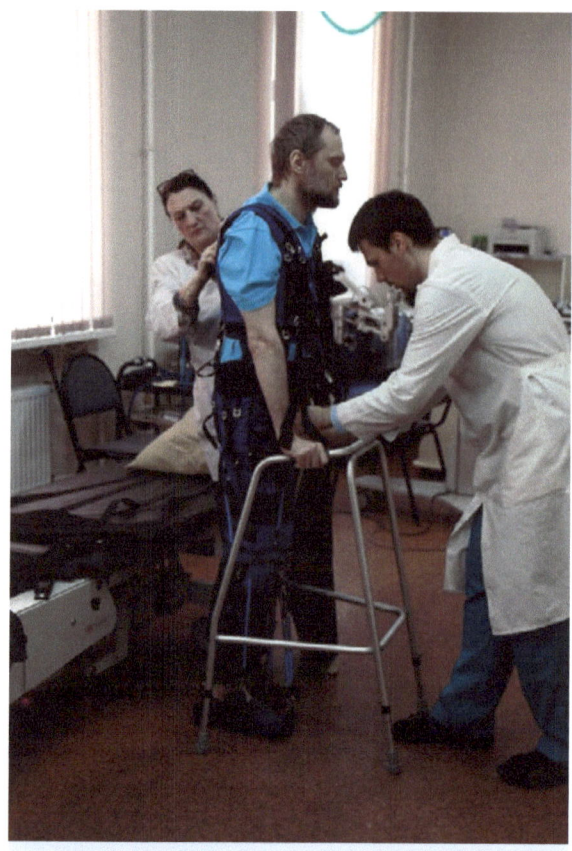

The Regent treatment suit in use for patient rehabilitation.

Image credit: Aerospace Medical Center and Technology, Russia

Center of Neurology under leadership of professor Ludmila Chernikova. These studies have shown that use of the suit results in a significant decrease in paresis and spasticity in the lower leg muscle groups, as well as an improvement of sensitivity in distal parts of lower limbs, and an overall improvement of locomotor functions. The positive effect on high mental functions was noticed at the same time, namely, an improvement of speech characteristics, an increase of active vocabulary, and an improvement in the patient's ability to recognize objects.

The use of the Regent suit is a complex, drug-free approach to the treatment of motor disorders. The method is closely related to the natural function of walking, activates all of the muscles involved in posture and spatial orientation and is very safe. It allows for shorter treatment time, can be used both under hospital and outpatient conditions and allows for a wide range of adjustments that allow individualized rehabilitation programs based on uniqueness of the neurological deficit and functional abilities of each patient. Today the Regent suit is applied in 43 medical institutions in Russia and abroad, and the results related to using both the Adeli and Regent suits are based on numerous observations and clinical studies.

Editor's note: *Small studies based in the U.S. and Israel were not able to distinguish improvements from Adeli suit therapy with traditional physical therapy. Therefore, this therapy has not yet been adopted in North America.*

High oblique view of an eruption plume from Kliuchevskoi Volcano on the Kamchatka Peninsula, Russia. The International Space Station was located over a ground position more than 1,500 km to the southwest when the image was taken on November 16, 2013.
Image credit: NASA

Earth Observation and Disaster Response

The International Space Station is a "global observation and diagnosis station." It promotes international Earth observations aimed at understanding and resolving the environmental issues of our home planet. A wide variety of Earth observation payloads can be attached to the exposed facilities on the station's exterior as well as in the Window Observational Research Facility located within the Destiny module. The presence of a human crew also provides a unique capability for real-time observation of the Earth, and "on the fly" data collection using hand-held digital cameras, and the astronauts may also provide input to ground personnel programming the station's automated Earth observation systems. Several instruments are currently collecting data from the International Space Station; in addition, some instruments have completed their data collection missions, with other remote sensing systems in development or proposed by researchers from the partner countries, NASA, academic institutions, and corporations. The existing international partnerships, fundamental to the International Space Station, facilitate data sharing that can benefit people around the world and promote international collaboration on other Earth observation activities. The station contributes to humanity by collecting data on the global climate, environmental change, and natural hazards using its unique complement of crew-operated and automated Earth observation payloads.

Environmental Earth Observations

The space station offers a unique vantage for observing the Earth's ecosystems and atmosphere with hands-on and automated equipment. The size, power, and data transfer capabilities of the station enable a wide range of sophisticated sensor systems including optical multispectral and hyperspectral imaging systems for examining the Earth's land surface and coastal oceans, as well as active radar and Light Detection and Ranging (LiDAR) systems useful for investigating sea surface winds and atmospheric aerosol transportation patterns. Astronauts using hand-held digital cameras provide an additional imaging capability for obtaining both detailed images of the Earth surface as well as sweeping panoramic views of its atmosphere. This flexibility is an advantage over sensors on unmanned spacecraft, especially when unexpected natural events such as volcanic eruptions and earthquakes occur.

Earth remote sensing from the space station

The installation of new facilities and sophisticated internal and external remote-sensing systems have transformed the International Space Station into a capable platform for Earth remote sensing. It also retains the unique distinction of being the only such platform with a human crew, which provides unique opportunities and advantages for remote sensing, particularly in the arena of data collection for disaster response efforts. So what can the station offer in terms of Earth remote sensing that free-flying, robotic satellite systems cannot?

> The space station significantly improves our ability to monitor the Earth and respond to natural hazards and catastrophes.

Images With a Variety of Lighting Conditions

Unlike many of the traditional Earth observation platforms, the station orbits the Earth in an inclined equatorial orbit that is not sun-synchronous. This means that the station passes over locations on the Earth between 52 degrees north and 52 degrees south latitude at different times of day and under varying illumination conditions. Robotic, satellite-based, Earth-observing sensors are typically placed on polar-orbiting, sun-synchronous platforms in orbits designed to pass over the same spot on the Earth's surface at approximately the same time of day.

Responsive Data Collection

Another advantage unique to the space station is the presence of crew that can react to unfolding events in real time, rather than needing a new data collection program uploaded from ground control.

Crew Earth Observations of forest fires near Sydney, Australia, taken just after local sunset with a handheld camera on Oct. 23, 2013. Orange fires and grey smoke threaten nearby towns.

Image credit: NASA

This is particularly important for collecting imagery of unexpected natural hazard events such as volcanic eruptions, earthquakes and tsunamis. The crew can also determine whether viewing conditions—like cloud cover or illumination—will allow useful data to be collected, as opposed to a robotic sensor that collects data automatically without regard to quality.

This is well demonstrated by the International Space Station (ISS) response to disaster events, in support of the International Charter, Space and Major Disasters (http://www.disasterscharter.org/home), also known as the International Disaster Charter (IDC). The ISS became a participating platform—in other words, a potential source of remotely sensed data—in April 2012, joining many other NASA satellite assets. As of May 2015, the NASA-managed sensor systems on ISS have responded to 130 IDC activations, with data collected for 34 of those events by either astronauts or ground-commanded sensors (or both).

The following NASA Earth observation instruments and facilities are now aboard and operational on the space station. International Partner sensor systems and programs are described in other articles included in this volume.

- Window Observational Research Facility (WORF; http://worf.msfc.nasa.gov/) provides a highly stable, internal mounting platform to hold cameras and sensors steady while offering power, command, data and cooling connections.

- ISS-RapidSCAT (http://www.jpl.nasa.gov/missions/iss-rapidscat/) monitors ocean surface wind speed and direction to provide essential measurements used in weather predictions, including hurricane monitoring. The sensor is mounted on the Columbus Module External Payload Facility and measures the echo strength of microwaves reflected off of the ocean surface. Several views of the same sea surface area provides radar return signals that can be used to estimate wind speed and direction.

- The Cloud Aerosol Transport System (CATS; http://cats.gsfc.nasa.gov/) is a LiDAR sensor that obtains measurements of atmospheric aerosols and clouds. Clouds and aerosols reflect a significant proportion of the sun's energy back to space, but their complex interaction in Earth's atmosphere is not yet fully understood. Data from CATS will allow scientists to better assess the role and impact of clouds and aerosols on Earth's global energy budget and climate.

- The High-Definition Earth Viewing (HDEV) camera (http://eol.jsc.nasa.gov/HDEV) includes four different commercial, high-definition cameras on the Columbus External Facility. The investigation is assessing camera quality while taking Earth imagery and the hardware's ability to survive and function in the extreme thermal and radioactive environment of low-Earth orbit.

- International Space Station SERVIR Environmental Research and Visualization System Pathfinder (IS-ERV; http://www.nasa.gov/mission_pages/station/research/experiments/867.html) is a sensor system collecting visible-wavelength imagery at ground resolutions of approximately 3 meters per pixel, which completed its mission in early 2015.

- Crew Earth Observations (CEO; http://eol.jsc.nasa.gov) includes Earth imagery taken by crew members using handheld cameras.

International Space Station SERVIR Environmental Research and Visualization System Pathfinder (ISERV) image of flooding in Cambodia, taken Nov. 1, 2013.

Image credit: NASA

The combined capabilities of both human-operated and autonomous sensor systems aboard the space station are helping to significantly improve our ability to monitor the Earth and respond to natural hazards and catastrophes. Integration of the space station Earth observation systems represents a significant and complementary addition to the international, satellite-based, Earth-observing "system of systems," providing knowledge and insight into our shared global environment.

Coastal ocean sensing extended mission

Scanning the globe from the vantage point of the International Space Station (ISS) is about more than the fantastic view. While cruising in low-Earth orbit on the ISS, the Hyperspectral Imager for the Coastal Ocean (HICO) gave researchers a valuable new way to view the coastal zone.

> Coastal waters are an important link between local and global economic development and environmental sustainability, and HICO on the space station was the first space instrument designed to observe them.

HICO, originally a one-year demonstration program supported by the Office of Naval Research (ONR) was installed on the ISS on Sept. 23, 2009, and collected data until Sept. 14, 2014. HICO far exceeded its planned mission and provided new insights into coastal environments around the world.

Using the HICO imaging spectrometer mounted outside the station on the Japanese Experiment Module - Exposed Facility of the Kibo Laboratory, researchers collected data about the Earth that will help them to better understand coastal environments and other regions around the world.

HICO was the first spaceborne, hyperspectral imager optimized for environmental characterization of the coastal ocean and large lakes. Archived HICO image data is being exploited to produce maps of coastal ocean properties, and provides a new science dataset for coastal regions worldwide. The HICO experimental sensor offered 10 times the spectral and spatial resolution of other ocean color sensors.

Why is this important? Coastal waters are an important link between local and global economic development and environmental sustainability. Coastal zones support many of the world's major cities (and their industrial zones, ports, and recreational facilities); they also include critical ecosystems that support fisheries and protect shorelines. Similarly, large lakes and reservoirs provide many of the same benefits and are the water supply for millions of people.

The ONR sponsored the development and first three years of the operation of HICO as an Innovative Naval Prototype. The HICO prototype had two goals. The first was to demonstrate ways to drastically reduce the cost and schedule of building a space payload. By innovative design and using commercial-off-the-shelf components where possible, the Naval Research Laboratory (NRL) engineers effectively built HICO in 18 months at a small fraction of the cost of traditional space instruments. The second goal was to demonstrate the ability to produce valuable images of coastal environmental properties using hyperspectral imagery from space.

Hyperspectral Imager for Coastal Ocean being installed on the Japanese Experiment Module - Exposed Facility of the Kibo Laboratory, Sept. 23, 2009.
Image credit: NASA

Hyperspectral Imager for Coastal Ocean image of a massive Microcystis *bloom in western Lake Erie, Sept. 3, 2011, as confirmed by spectral analysis.*
Image credit: NASA NRL OSU

These goals were met and the mission extended by NASA continuing to fund operation of the sensor. HICO data was used by scientists at NRL, Oregon State University and over 50 other institutions worldwide to study coastal environments. Access to the instrument by researchers in U.S. government agencies, commercial entities and other non-academic institutions was also available through proposals to the Center for the Advancement of Science in Space (http://www.iss-casis.org). Studies ranged from characterizing coral reefs of Australia, New Caledonia, and Palau to assessing water quality in lakes and reservoirs in North America and Europe to developing algorithms for NASA and European-proposed, hyperspectral instruments.

The results from HICO investigations provide benefits to agencies with marine responsibilities, such as the National Oceanic and Atmospheric Administration (NOAA) via information on bathymetry, bottom type, water clarity, and other water optical properties. Scientists at the Environmental Protection Agency used HICO data (http://www.nasa.gov/mission_pages/station/research/news/epa_coastal.html) to develop water quality monitoring tools that will allow the public to check water conditions from their mobile devices. The rich hyperspectral information available from HICO enabled the use of spectral techniques to specifically identify features such as the massive *Microcystis* bloom in Western Lake Erie (shown above), which threatened the water supply for millions of people in September 2011.

HICO was operated under ONR sponsorship for over three years and in 2013, NASA assumed sponsorship of operations in order to leverage HICO's ability to address their Earth-monitoring mission. This opened up access of HICO data to the broad research community. All the HICO data is now available on NASA's Ocean Color website (http://oceancolor.gsfc.nasa.gov/).

Visual and instrumental scientific observation of the ocean from space

One feature of oceanographic research conducted with the participation of cosmonauts on orbital stations, including the historical Salyut, Mir, and the current International Space Station (ISS), is the broad application of the method of scientific visual and instrumental observation of the world's oceans from space. The basis of the method is the visual search, detection and identification of phenomena under examination in the near-surface layer of the ocean and the atmosphere above it. This is the simplest, yet one of the most informative, ways to obtain data in the visible spectrum on the condition of the ocean's natural environment. The reliability and scientific value of information on the ocean obtained in this way significantly increases because of the targeted use of special recording equipment (such as digital photo and digital video cameras) and of on-board instruments that expand the capabilities of the crew member's visual analysis capabilities during observations. Such combined observations are referred to as visual and instrumental. The methods of visual and instrumental

> Visual and instrumental observations of the ocean from space have broad practical applications for resolving the issues pertaining to the research of biological resources of the world's oceans.

observation (VIO methods) are used to establish informational databases in the visible electromagnetic wave spectrum not only in the field of remote sensing of the oceans but also in other areas of knowledge and practical activity.

The many years of Russian experience conducting oceanographic experiments by crews on the long-term, orbital stations, Salyut, Mir, and ISS, have made it possible to evaluate the actual informational potential of VIO of the world's oceans from space. Also, this has allowed for the development of a flight-tested method of solving specific issues of oceanography and of developing equipment and procedures for the remote sensing of the ocean. These procedures and methods are frequently reviewed while conducting oceanographic experiments with cosmonauts' participation, as is currently done on the ISS as part of some space experiments (e.g. "Diatomea," "Seiner," etc.).

The main object of search and observation for a cosmonaut researcher while working on this category of task using VIO is large-scale, color-contrast formations (CCF) on ocean surfaces related to the mass growth of phytoplankton (Fig. 1). In the field of view of space station crews observing the ocean surface along the flight path, a wide variety of cloud formations are constantly present. In addition, the following are observed among cloud fields above the ocean: cloud indicators of tropical cyclones in varying stages of development (Fig. 2), lineaments (Fig. 3) identifying jet streams, cumulus clouds with powerful vertical development above the ocean surface under intensive atmospheric convection (Fig. 4), and other phenomena of interest for maritime meteorology serving the shipping industry, aviation, and seafood industry.

Of particular importance among hydro-meteorological phenomena observable by the VIO method are tropospheric cloud formations characteristic for the movement of air masses past obstacles in the atmosphere of island regions (Karman vortex streets, Helmholtz gravitational shear waves, etc.). Experiments have also proven the capability to identify and record, through imagery from space, optically active events in the atmosphere, such as terrigenous dust and sand flows (Fig. 13), fog, volcanic ash clouds over the ocean, etc., and regions with signs of intensive thunderstorm activity. Applicable to the hydro-physical area of oceanic research, the VIO method ensures obtaining documented data on the nature of local water circulation (Fig. 1), icebergs (Fig. 6), the structure of surface agitation fields (Figs. 7 and 8), broken ice (Fig. 5), and the color and transparency of water (Figs. 1, 9 and 10). The results of such data interpretation are used to describe the most significant elements of general ocean water circulation and to deal with hydro-optic tasks.

To date, a significant amount of information obtained from the visual and instrumental observations of the oceans from space, has been collected and grouped according to various objects of environmental monitoring. As applicable to open ocean ecosystems, the most broadly represented are the results of observations and color photographs of the ocean characterizing the diversity of forms and condition of coral reefs (Figs. 11 and 12), the morphology of different sizes of phytoplankton fields and the hydro-dynamic specifics of the environments in which they live (Fig. 1).

The most important aspect of the VIO method is the capability to evaluate the environmental condition of the ocean-atmosphere system in real time, to identify anomalous processes and phenomena in the ocean environment, such as surfactant films (Fig. 14), oil and petroleum products spills (Fig. 16), contamination of clean ocean waters with surface runoff (Fig. 15), and rinsing agents of ferromanganese nodules mined from the sea floor in mining areas of the oceans.

Currently, visual and instrumental observations of the ocean from space have broad practical applications for resolving the issues pertaining to the research of biological resources of the world's oceans. The increased attention to this area of research is explained by the relevance of the problem, by the capability to conduct research using relatively inexpensive commercial photography equipment, and by the existence of an algorithm of searching and identifying from space, highly productive waters of the world's oceans that have been tested by crews on Russian space stations and patented.

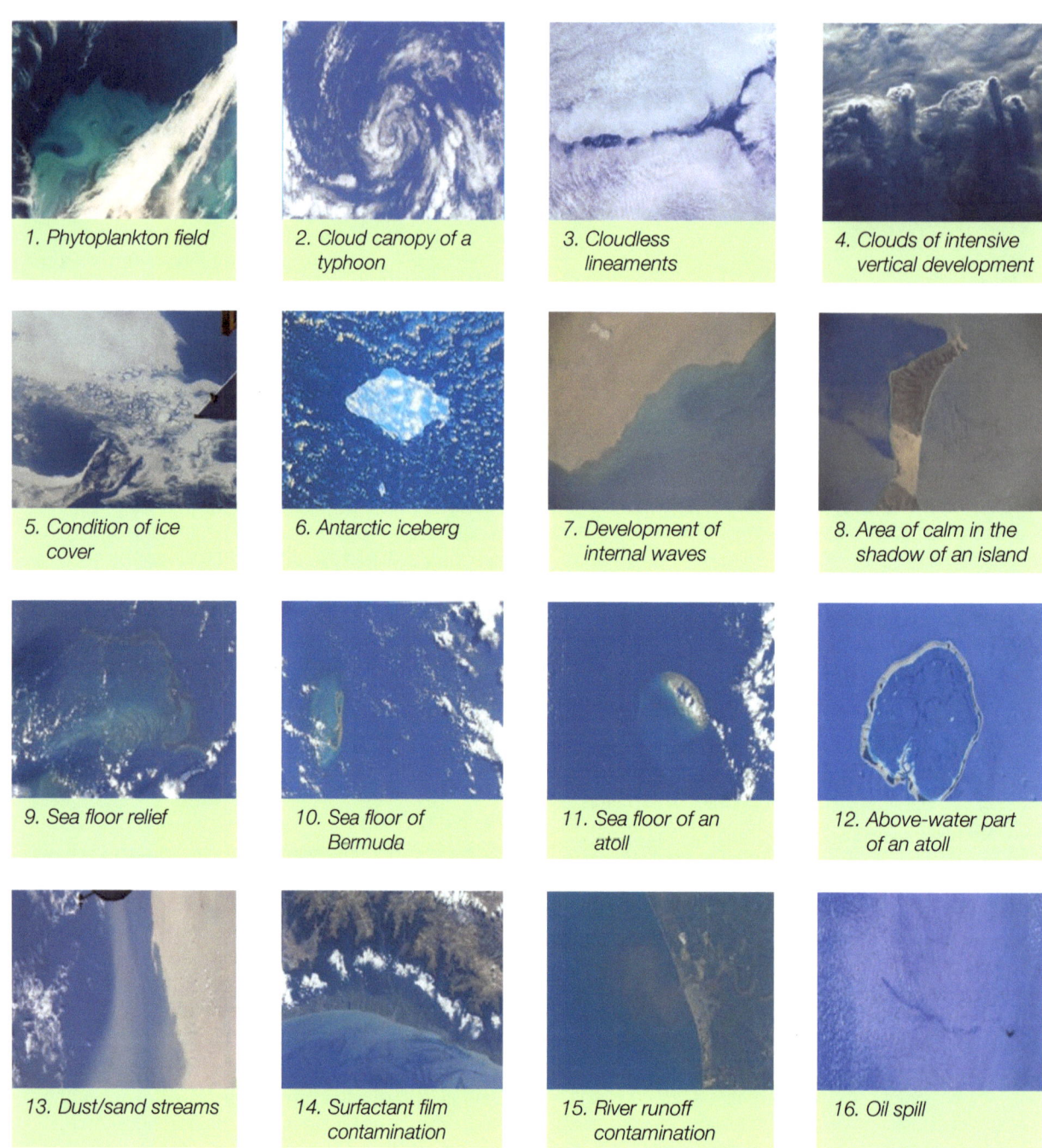

Image credit: Roscosmos/Energia/FGUP TsNIIMash

Disaster Response

Remotely sensed data acquired by orbital sensor systems has emerged as a vital tool to identify the extent of damage resulting from a natural disaster, as well as providing near-real time mapping support to response efforts on the ground and humanitarian aid efforts. The International Space Station (ISS) is a unique terrestrial remote-sensing platform for acquiring disaster-response imagery. Unlike automated remote-sensing platforms it has a human crew; is equipped with both internal and externally mounted still and video imaging systems; and has an inclined, low-Earth orbit that provides variable views and lighting (day and night) over 95 percent of the inhabited surface of the Earth. As such, it provides a useful complement to autonomous sensor systems in higher-altitude polar orbits for collecting imagery in support of disaster response.

Space station camera captures Earthly disaster scenes

Of all the hundreds of spacecraft and satellites in low-Earth orbit, few can match the International Space Station (ISS) for its view of the big, blue marble below it. The ISS, circling the Earth once every 90 minutes, offers a unique observing platform with over 90 percent of the Earth's populated area visible from its orbit. Scientists put that spectacular view to good use by using a camera on the ISS to demonstrate disaster observation and responses to humanitarian requests.

The Royal Gorge wildfire ignited on June 11, 2013, northwest of Cañon City and along the Arkansas River in central Colorado. The fire raged over the sagebrush and pine-covered topography, charring 3,218 acres in five days until firefighters finally corralled it. This ISS SERVIR Environmental Research and Visualization System (ISERV) image shows the Arkansas River passing through the burn scar.

Image credit: NASA

> The ISERV camera helped the space station support countries around the world, making the ISS even more of an international asset by capitalizing on the unique view of Earth it provides.

The system is a prototype called the ISS SERVIR Environmental Research and Visualization System (ISERV) Pathfinder. ISERV was developed by NASA to support a joint NASA/US Agency for International Development (USAID) project known as SERVIR (http://www.nasa.gov/mission_pages/servir/index.html) and potentially

June floods devastated much of southern Alberta, Canada, and forced the evacuation of over 100,000 citizens in Calgary and nearby towns. Three people died in the swirling, murky waters, which also caused millions of dollars' worth of damage on June 22 and the days following. Before and after images of flooding in downtown Calgary are shown here. The GeoEye/IKONOS (a commercial satellite sensor) image on the left, taken on Sept. 13, 2008, shows normal flow of the Bow and Elbow Rivers. The ISERV image, captured June 22, 2013, is on the right and shows floodwaters from the two rivers inundating downtown Calgary. Canadian officials used the images to help in their disaster assessments and to improve their flood-mapping algorithms.

Image credits: left, Digital Globe; right, NASA

the broader NASA Applied Sciences community. The word "servir" is Spanish for "to serve." The SERVIR project provides satellite data and tools to environmental decision makers in developing countries and operates via regional "hubs" in Nairobi, Kenya; Kathmandu, Nepal; and Panama City, Panama. These SERVIR hubs can task the ISERV system to image scenes of Earth's surface in their countries to address environmental issues and disasters. SERVIR's coordination office at NASA's Marshall Space Flight Center in Huntsville, Alabama, controls ISERV operations.

Installed in January 2013 for a two-year mission, the camera observes our planet's surface through the Destiny module's Earth-facing window, acting on commands to photograph specific areas during the space station's next pass over them. ISERV's targets were threatened by or already experiencing floods, landslides, forest fires, or other disasters. The images were used to monitor the situation, evaluate damage extent and direct evacuation and disaster relief efforts. ISERV was well adapted to higher-resolution "survey" applications like disaster monitoring and assessment.

ISERV captured first light on Feb. 16, 2013, and the myriad images it has captured since then of locales around the world include critical disaster photos. As two examples, ISERV images revealed the burn scar left by the June 11-16, 2013, Royal Gorge wildfire in

Colorado, and the devastation caused by floodwaters in Calgary on June 22-24, 2013.

While the ISERV camera completed its primary mission goals in December 2014, the system remains aboard the International Space Station and could be re-deployed if needed to capture imagery of natural disasters. The ISERV camera can help the space station lend support to countries around the world, making the ISS even more of an international asset by capitalizing on the unique view of Earth it provides.

Editor's note:

Georeferenced images are available for public use at: ftp://ghrc.nsstc.nasa.gov/pub/iserv/data/L0/.

The SERVIR team provides an online map/tool (http://www.servirglobal.net/mapresources/iserv/) that allows users to locate and download ISERV images.

Clear high-definition images aid disaster response

Data collected from various International Space Station (ISS) sensor systems have contributed to Earth observation and disaster response through international collaboration frameworks, such as the International Charter, Space and Major Disasters (http://www.disasterscharter.org/home) and Sentinel Asia (http://www.jaxa.jp/article/special/sentinel_asia/index_e.html). The Japanese Experiment Module (JEM), or Kibo, provides opportunities to obtain very clear high-definition (HD) images both from internal handheld and externally mounted cameras. These clear images are beneficial for disaster support.

The Japan Aerospace Exploration Agency (JAXA) offers data taken with two camera systems, the Super Sensitive HDTV Camera System (SS-HDTV) and the Commercial Off-The-Shelf (COTS) High-Definition Television Camera System on JEM External Facility (HDTV-EF). JEM-EF is an unpressurized, multipurpose pallet structure attached to the JEM. This external platform is used for research in diverse areas such as communications, space science, engineering, technology demonstration, materials processing, and Earth observation.

SS-HDTV was developed to take night images of the Earth, including such phenomena as aurora, airglow and meteor showers. It is operated in the ISS pressurized module cabin including the JEM and the Cupola Observational Module. The beautiful night images are utilized for the check of the electric power restoration and the revival of cities after a natural disaster and the return of normal life to those people affected.

Images of the Earth surface, the ocean, clouds, etc., are taken from ISS for disaster response, education and publicity purposes.

> Data collected from various ISS sensor systems have contributed to Earth observation and disaster response through international collaboration frameworks.

Night view of Italy.
Image credit: JAXA/NASA

Night view, aurora and airglow.
Image credit: JAXA/NASA

Wildfire, Queensland, Australia.
Image credit: JAXA/NASA

Hurricane Sandy.
Image credit: JAXA/NASA

Sentinel Asia aims to promote international cooperation to monitor natural disasters in the Asia-Pacific region. According to statistics by the Asian Disaster Reduction Center's Natural Disasters Data Book 2013, Asia accounts for 44.6 percent of occurrences; 84.6 percent of people killed; 87.1 percent of affected people; and 49.0 percent of economic damage. Under these circumstances, the Asia-Pacific Regional Space Agency Forum (APRSAF) proposed Sentinel Asia in 2005 to showcase the value and impact of Earth observation technologies.

Sentinel Asia uses Earth observation satellites and other space technologies to collect disaster-related information and shares it over the internet. The aim is to mitigate and prevent damage caused by natural disasters such as typhoons, floods, earthquakes, tsunamis, volcanic eruptions, and wildfires. Sentinel Asia now counts 15 international organizations and 83 participating organizations from 25 countries as members, and utilization of its systems is steadily expanding. JAXA, as the only Asian partner of the ISS, will continue to support disaster response and hopes to contribute to Asia and the whole world with Kibo and its high-definition cameras.

Water filtration plant set up in Balakot, Pakistan, following the earthquake disaster in 2005. The unit is based on space station technology and processes water using gravity fed from a mountain stream.
Image credit: Water Security™ Corporation

Innovative Technology

In space, physical processes can be better understood with the control of external influences such as gravity. Technical innovations designed for space systems are tested on the International Space Station (ISS) before use in other spacecraft systems. While investigating how new technologies operate in space, unexpected discoveries are possible. Simplified physical systems can also be directly used to improve models of physical processes leading to new industrial techniques and materials.

The ISS provides the unique capability to perform long-duration experiments in the absence of gravity and in interaction with other spacecraft systems not available in any other laboratory. Additional insight comes from the presence of the ISS crew observing and interacting with these experiments and participating in the discovery process.

The ISS research portfolio includes many engineering and technology investigations designed to take advantage of these opportunities. Experiments investigating thermal processes, nanostructures, fluids and other physical characteristics are taking place to develop these technologies and provide new innovations in those fields. Additionally, advanced engineering activities operating in the space station infrastructure are proving next-generation space systems to increase capabilities and decrease risks to future missions. Emerging materials, technology and engineering research activities on the ISS are developing into benefits for economic development and quality of life.

Fluids and Clean Water

Whether in the vacuum of space or the relative comfort of the Earth's surface, access to clean water is essential for living organisms. The challenges of moving and processing fluids such as water using compact, reliable systems in the microgravity environment of space have led to advances in the way we purify water sources on the ground. Testing methods developed to ensure water quality on the International Space Station (ISS) have led to advancements in water monitoring here on Earth. Investigations into the basic dynamics of how fluids move in space have also led to advances in medical diagnostic devices.

Advanced ISS technology supports water purification efforts worldwide

Whether in the confines of the International Space Station or a tiny hut village in sub-Saharan Africa, drinkable water is vital for human survival. Unfortunately, many people around the world lack access to clean water. Using technology developed for the space station, at-risk areas can gain access to advanced water filtration and purification systems, making a life-saving difference in these communities.

> Using technology developed for the space station, at-risk areas can gain access to advanced water filtration and purification systems, making a life-saving difference in these communities.

In 2006, the first of many ground-based water filtration systems using NASA technology was installed in northern Iraq. The system was developed by Water Security Corporation in Reno, Nevada, and installed by the nonprofit organization Concern for Kids (CFK). CFK representatives learned about a deep-water well failure in the tiny Kurdish village of Kendala, Iraq, which left its residents without access to drinkable water. The population quickly dwindled from 1,000 residents to a mere 150. Those remaining were forced to use a nearby creek that contained water contaminated by livestock, which they sifted through fabric to remove dirt and debris.

Todd Harrison was president of CFK's board of directors at the time and strongly empathized with the people of Kendala. He set out on a mission to revive the ailing community by improving the deplorable conditions. The solution came in the form of a familial connection that put Harrison in touch with NASA engineers who developed technology to provide clean water aboard the space station.

Harrison's sister, Robyn Gatens, was the engineering manager for the Environmental Control and Life Support System (ECLSS) project at NASA's Marshall

Girl at hydration station.
Image credit: Sinergia Systemas

Space Flight Center. She and her team of engineers were responsible for developing the cutting-edge water purification system that recycles air and water aboard the station.

By efficiently recycling wastewater aboard the space station, there is a reduced need to provide the resource via resupply—which would not be an option for long-duration space travel. Without this capability, the station's current logistics resupply capacity would not be able to support the standard population of six crew members.

Two principal components make up the International Space Station Regenerative ECLSS: The Water Recovery System (WRS) and the Oxygen Generation System (OGS). The WRS conducts the water purification and filtration process in the ECLSS. Water Security Corporation (WSC) took an interest in this part of the ECLSS project, and licensed the technology in order to adapt it to an Earth-based water treatment system.

Harrison discovered an interesting relationship between WSC's water filtration system and NASA because of his familiarity with his sister's work. NASA's previous research and application provided the Microbial Check Valve (MCV), an integral component of the purification and filtration process.

The MCV is an iodinated-resin that provides a simple way to control microbial growth in water without the use of power. By dispensing iodine into the water, it performs an important secondary nutritional function for the populace. When added to the diet, it promotes proper brain function and maintains bodily hormone levels, which regulate cell development and growth. Children born in iodine-deficient areas are at risk of neurological disorders and mental retardation.

With the help of U.S. Army Civil Affairs and Psychological Operations Command (Airborne) personnel, a 2,000-liter water tank and fresh water were delivered to the Kurdish village in Iraq. Workers ensured that the water was clean and iodinated to prevent bacteria and virus contamination. When CFK encountered technical issues, Gatens and her team were able to help by phone to implement a workaround that enabled the successful processing of Kendala's water supply.

Joint collaborations between aid organizations and NASA technology show just how effectively space research can adapt to contribute answers to global problems. Since this initial effort, the commercialization of this station-related technology has provided aid and disaster relief for communities worldwide. WSC, in collaboration with other organizations, has deployed systems using NASA water-processing technology

Chiapas installed system at school.
Image credit: Sinergia Systemas

around the world. Applications have included home water purifiers in India, village processing systems in remote areas of Mexico, Central and South America, water bottle filling stations in Pakistan, and even a survival bag designed as a first response device for natural disasters, refugee camps, civil emergencies and remote locations.

Exploring the wonders of fluid motion: Improving life on Earth through understanding the nature of Marangoni convection

Fluid is everywhere in our lives. The Earth, known as "the water planet," is able to support life in part because of the presence of water. From the lava that

> ISS gives us the unique opportunity to study a fundamental principle of motion, Marangoni convection, which is revealed in microgravity but masked by gravity on Earth.

cools to form islands, the blood that flows in our veins, and the molten metals that we turn into structures and vehicles, human beings have been using fluids throughout history. Understanding the fundamental principles of fluid motion is important for all walks of life, from the microfluidic systems that deliver drugs to keep us healthy to the rocket fuel tanks that propel us into the vastness of space. One of these fundamental principles of motion, Marangoni convection, is revealed in microgravity but masked by the stronger force of gravity on Earth. The International Space Station gives us the unique opportunity to study this principle in detail.

Marangoni convection is the flow driven by the presence of a surface tension gradient that can be produced by a temperature difference at a liquid/gas interface. It can best be studied in a liquid bridge formed under microgravity conditions. To learn thermal-fluid dynamics in microgravity, the convection in a liquid bridge of silicone oil is generated by heating one disc higher than the other, allowing scientists to observe flow patterns that can tell them about how heat is transferred in microgravity.

Surface tension is the characteristic of a liquid in which it forms a layer at its surface so that this surface covers as small an area as possible. For example, in the image below, one can see the coin floating on the surface of the water. Surface tension is the force that keeps the heavier coin from sinking. In general, surface tension becomes stronger with decreasing temperature. A

The Earth, known as "the water planet."
Image credit: JAXA/Japan Broadcasting Corporation

A coin floats on the surface of the water because of surface tension.
Image credit: JAXA

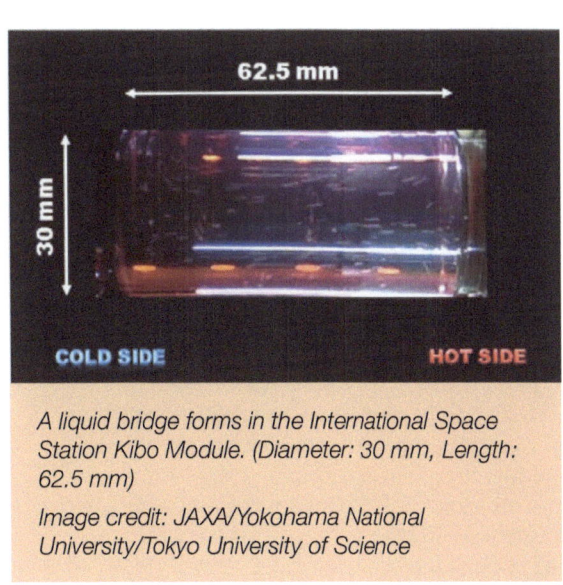

A liquid bridge forms in the International Space Station Kibo Module. (Diameter: 30 mm, Length: 62.5 mm)
Image credit: JAXA/Yokohama National University/Tokyo University of Science

Pictured are the tears or legs of wine. Droplets form the vicinity of the fluid's surface and drip into the wine continuously.

Image credit: Professor Hiroshi Kawamura, Tokyo University of Science

trait of Marangoni convection is a surface tension-driven fluid flow in which the driving force is localized only at the surface. When a temperature difference exists along a surface, the surface is pulled toward a low-temperature region. The surface tension difference is also produced under fluid concentration differences at the meniscus (the fluid's surface), such as what you may see in a wine glass as tears or legs of wine.

Such a phenomenon is often observed in everyday life. For example, oil in a pan heated from the center moves to the side. Oil floating on water immediately moves when a surfactant (e.g., detergent) drops onto a part of the oil because of the imbalance in the surface tension. The detergent causes the center to have a lower surface tension while the outside has a higher surface tension; therefore, the center and the oil were pulled out in all directions to equalize the surface tension. These phenomena result from the Marangoni effect.

The Marangoni experiment has obtained the flow transition process from laminar to turbulent (chaotic) flows. The onset conditions of oscillatory flow were clarified by studying long liquid bridges, which are only available under the microgravity environment present on the space station. This helps to make more accurate predictions of instability onsets that can give rise to different pattern-forming instabilities. In a higher temperature difference regime, transitions to the chaotic and turbulent flows are detected and characterized. This knowledge of the floating zone (liquid bridge), may help in refining a technique that is used to produce extremely high purity crystals by the semiconductor and rare metals industries. Flow disturbance is a major cause for the deterioration of the quality of the crystal grown by this method.

Understanding the rheological (deformation and flow of matter) dynamics in liquid bridges is of fundamental interest for many industrial and biological processes, as well as medical diagnostic devices. For instance, micro-Total Analysis Systems (μ-TAS) are expected to enable on-site medical diagnosis techniques by sampling minute amounts of blood and DNA. In this technique, fluid manipulation is dominated by the interface capillarity. With a stronger understanding of the fundamentals of Marangoni phenomena, practical use of thermocapillarity in microfluidics such as the μ-TAS can be achieved.

Space station-inspired mWater app identifies healthy water sources

What if that clear, sparkling stream coming from the ground or a faucet were teeming with contaminants? How would you know? Whether you live in some remote region of Africa, a high rise in New York City or aboard an orbiting laboratory in space, you need reliable drinking water to survive. You now can check the cleanliness of your water using the mWater app on your mobile phone.

This handy tool, based in part on International Space Station technology, provides a global resource available for free download as an app or usable via the Web browser version of the app on most smartphones. Governments, health workers and the public all can make use of mWater to record and share water test results. During the first year of the beta release of mWater, more than 1,000 users downloaded it and mapped several thousand water sources.

John Feighery, mWater co-founder and former lead engineer for air and water monitoring with NASA, was

> Check the cleanliness of your local water bodies using the mWater app first developed from technology proven on the space station.

Screenshot of the global water source map people can visit to find data uploaded by mWater users across the globe.

Image credit: mWater

inspired by his work for the space station. There, he and his team created efficient, mobile and ambient testing techniques to test for contamination in drinking water sources without the need for costly lab equipment such as incubators. The resulting Microbial Water Analysis Kit (MWAK)—part of the environmental monitoring Crew Health Care System Environmental Health System (CHeCS EHS) suite aboard station—sparked Feighery's imagination, providing the basis for the mWater testing of E. coli in 100-milliliter (3.38-ounce) water samples.

One key innovation that came from NASA was proving that these types of tests will work at near ambient temperatures. Various studies have shown that any temperature about 25 degrees Celsius (77 degrees Fahrenheit) will produce a result, whereas traditional laboratory procedures call for incubation at 37 degrees Celsius (98.6 degrees Fahrenheit). This is very important for developing countries because incubators are expensive and require reliable electricity and can also easily break down. Since many of the countries that suffer from poor access to safe water are tropical, the tests can easily be done by anyone at room temperature most any time of the year.

Hefting testing materials or expensive equipment to test water sources is unrealistic, Feighery discovered while volunteering with Engineers Without Borders in El Salvador. Portable, inexpensive and effective, that's the goal for technologies bound for the defined real estate of the space station, but also for those needed in remote or low-resource regions of the world. Low-cost mWater test supplies cost users $5 per kit.

Combining his aerospace experience and philanthropic passions, Feighery went to work with co-founders Annie Feighery and Clayton Grassick on what would become mWater. Following the 2011 Water Hackthons, the mWater app was developed and later improved upon through field testing sponsored by U.N. Habitat. The app helps to simplify recording of water quality results, mapping water sources and finding safe water nearby.

The tests and app are both designed with ease of use in mind. The user tests the water, allows the test to incubate at ambient temperatures, photographs the results to count the bacteria and finally uploads the findings to the global water database. Ease of use was key to Feighery's design goals.

Feighery's experience with writing crew procedures at NASA influenced the design of the app. The app is task-oriented and designed to require very little training beyond following the procedure. In the future, Feighery and his mWater team plan to introduce checklists for each type of water test to further improve ease of use and reduce the training needed to perform field testing.

Test results upload to the cloud-based global water database, using the phone's Global Positioning System to identify the exact location of the water source. Each location gets a unique and permanent numeric identifier for reference by those who visit the global water source map for updates. Users can add new water location points and input or update test results, working within the open source sharing approach for the health of the community.

Space-tested fluid flow concept advances infectious disease diagnoses

A low-energy medical device that can diagnose infectious diseases on-site may soon be operating in remote areas of the world that have limited access to power sources. With a reduced need for energy and on-site diagnosis, less time would be needed between identifying a disease and beginning treatment for it.

The device that could quickly identify diseases like HIV/AIDS or tuberculosis relies on a deeper understanding of capillary flow. That deeper understanding is the result of research conducted on the International Space Station.

The Capillary Flow Experiment (CFE) was a suite of fluid physics experiments conducted on the space station by Dr. Mark Weislogel of Portland State

> Understanding capillary flow could change how fluid-handling systems are designed and operated in any number of applications.

University in Oregon and assisted by researchers at NASA's Glenn Research Center in Cleveland, Ohio.

Capillary flow, also known as wicking, is the ability of a liquid to flow without the assistance of gravity and other external forces. It even works in opposition to those forces. Stick a straw into a glass of water; the water will rise maybe a few millimeters in the straw before you begin to drink through it. Or consider how a paper towel will draw, or wick, liquid into it. In the absence of gravity, the effect of capillary forces is more dramatic. For example, the water would rise and completely fill a straw before you begin to drink through it.

CFE was a basic physics investigation that refined our understanding of how capillary action helps fluids flow. The principle has application in many fluid-handling systems from fuel tanks to cooling systems to medical devices.

Cell samples in the form of bodily fluids or blood are placed in the CFE-studied medical device. Enzymes burst the samples leaving behind DNA or RNA, which is then captured on a bead that is processed by the device to identify the infectious virus. Capillary flow is used to manage and direct the flow of the cell samples inside the device.

David Kelso, Ph.D., of Northwestern University in Evanston, Illinois, developed the simple, inexpensive device. Kelso and his team were using energy-consuming items like batteries and motors to operate the device, but when his designs did not work as expected in the lab, Kelso turned to Weislogel. Kelso explained that he and his team thought that gravity would pull fluids through the device, but Weislogel had the understanding that capillary action would do this based on his previous work in microgravity.

By relying on the principles of capillary flow, the device uses much less energy and can provide medical professionals with a valuable tool in areas with limited resources. The device is scheduled for field testing in late 2014.

While a primary focus of the CFE research was fluid management in space where gravity is nearly absent, the basic principles of capillary flow can be used on Earth as well. The most direct applications for CFE research are immediate design improvements for most life support equipment aboard spacecraft. In addition, Weislogel believes that terrestrial applications will be commercially viable, applying the unique results of space station research.

Research in the microgravity environment of space is once again contributing to work that impacts our lives here on Earth. Much like it is with the medical device, this deeper understanding of capillary flow could change how fluid-handling systems are designed and operated in any number of applications.

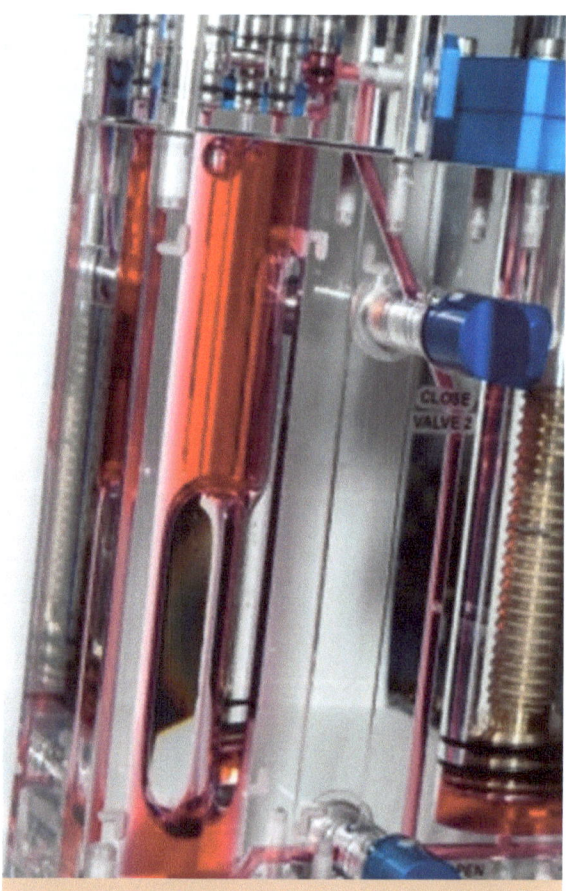

A view of Capillary Flow Experiment - 2 Interior Corner Flow 3 vessels set up during an experiment aboard the International Space Station. This study looks at capillary flow in interior corners.
Image credit: NASA

Materials

The ISS provides a unique laboratory environment for the testing of new materials. In microgravity, sedimentation and buoyancy-driven convection do not take place allowing us to witness how materials change and develop over longer periods. This allows researchers to manipulate their materials in unique ways. These opportunities are leading to a better understanding of how material processes work on Earth thereby enabling the manufacturing of new materials with well-defined structures, improved strength, and better function.

Improving semiconductors with nanofibers

Nanotechnology involves materials at the atomic and molecular level and holds great promise for a wide range of applications, from telecommunications and computing to health and medicine. But nano-structured materials, especially those that can self-assemble into organized patterns, have proven difficult to control. Experiments on the International Space Station (ISS) demonstrated a new process for constructing materials that can arrange themselves into structures just one atom thick (the nano scale). The research paves the way for development of fast, high-capacity computers and information storage systems.

> Data from ISS for nano-scale material assembly will be applied in several fields, including computers and chemical catalysts for industrial processes.

The 2-D Nano Template experiments were designed to build a fine layer of material that can assemble by itself into a very tight, repeating pattern. This pattern can be used as a template for a focused beam of electrons. The electron beam traces the template, carving out the same pattern onto another surface.

The investigation used peptides, which are small biological molecules made from amino acids, and polymers, which are larger molecules made from many smaller parts that repeat. Proteins are made up of peptides, and DNA is an example of a polymer. Binding together with water, the molecules would form nanofibers, which would combine to create a single array just one molecule thick. This ultra-thin surface would be used as a masking layer for a tiny semiconductor—the basis of a computer chip.

Investigators developed a simple way to induce the peptides to bond, starting the process of self-assembly. Microgravity provided a unique environment for this process, because it was not interrupted by the force of Earth's gravity causing some of the molecules to settle or clump together incorrectly.

After the samples were returned to Earth, researchers studied them using a special microscope called an Atomic Force Microscope (AFM), which enables a view of the atomic scale. The images showed that fibers built on the ISS were greatly improved compared to a sample grown on Earth. The space-based sample also had a much more uniform pattern because there were no excess molecules or particles settling together.

The new, error-free, two-dimensional nanofiber layer was then used as a template, which can be traced like a blueprint to mark the processing surface of a semiconductor. Investigators tested this in space with a silicon carbide substrate. When they coated it with the two-dimensional nano-patterned template, a focused ion-beam followed the pattern and marked it onto the silicon carbide material. This is a novel process for creating a very specific, incredibly tight pattern with less than 10 nanometers of spacing.

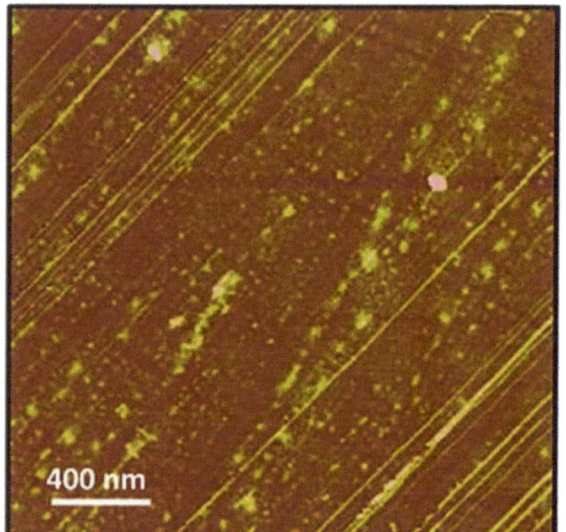 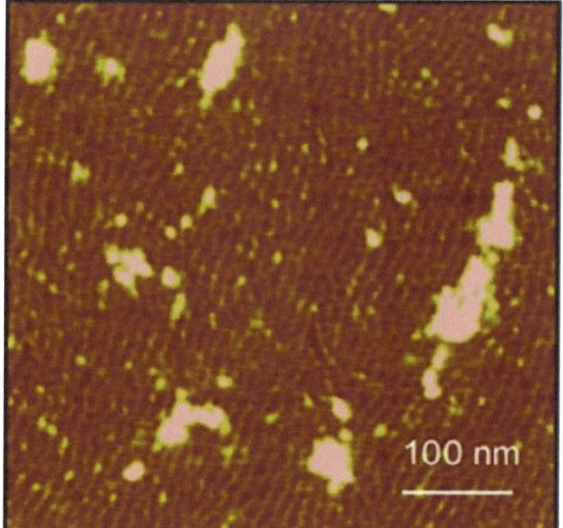

AFM images of (left) long-range fiber obtained in the space experiment and (right) original fiber array obtained in the ground experiment.
Image credit: JAXA/Nagoya Institute of Technology

Materials that can follow such a small-scale pattern will be useful in several fields, including computers, chemical catalysts for industrial processes, and even super-water repellent substances. Thanks to space station research, the cutting edge of nanotechnology is sharper than ever.

InSPACE's big news in the nano world

A technology of tiny elements studied on the International Space Station could have a big impact on everything from braking systems and robotics to earthquake-resistant bridges and buildings.

Investigating the Structure of Paramagnetic Aggregates from Colloidal Emulsions (InSPACE) is a set of experiments that gathered fundamental data about Magnetorheological (MR) fluids. They are a type of smart fluid that tends to self-assemble into shapes when exposed to magnetic fields.

MR fluids change viscosity in a magnetic field and can even be made to change their arrangement at the nanoscale level, or one billionth of a meter. Such tiny distances are typical for molecules and atoms.

When exposed to magnetic fields, MR fluids can quickly transition into a nearly solid state. When the magnetic field is removed, the MR fluids return to a liquid state. This process produces useful viscoelastic properties that can be harnessed for a variety of mechanical devices, from robotic motions to strong braking and clutch mechanisms. The process of self-organization exhibited by MR fluids also could have long-ranging consequences for the design and manufacturing of a whole host of new nanomaterials and nanotechnologies.

The Colloid Self Assembly set of experiments conducted during InSPACE-3 looked at colloid arrangement at a nanoscale using magnetic and electric fields for development of nanomaterials. The principal investigator for the study was Eric M. Furst,

> Magnetorheological fluid research in space could have long-ranging consequences for the design and manufacturing of a whole host of new nanomaterials and nanotechnologies.

Ph.D., University of Delaware. His study goal was to understand the fundamental science around directed self-assembly to better define new methods of manufacturing materials composed of small colloidal or nanoparticle building blocks.

Colloids are tiny particles suspended in a solution. They are critical to industrial processes as well as household products such as lotions, medications and detergents.

InSPACE-3 is focused on oval- or ellipsoid-shaped particles, as opposed to earlier InSPACE investigations with MR fluids composed of round particles. These oval- and ellipsoid-shaped particles were expected to pack differently and form column-like structures in a unique way, different from previous experiments. Particles of InSPACE-3 are made of a polystyrene material embedded with tiny, nano-sized iron oxide particles.

When the fluid containing iron oxide is mixed, it has a brownish, rusty hue. Astronauts, under the direction of the project team, ran a series of experiments on this rust-colored mixture.

Astronauts applied a magnetic field, which was pulsed from a low frequency of around 0.66 hertz up to 20 hertz, or switched on and off, roughly one time per second up to 20 times per second. Scientists were looking for formation of structures that are at a lower-energy state. Typically in an MR fluid application, a constant field is applied, and the particles form a gel-like structure. They don't pack very well, so the particles have no definite form. They are like a cloud or hot glass that can form into almost any shape.

In a pulsed field, the on-off magnetic field forces the particles to assemble, disassemble, assemble, disassemble and so on. In this pulsed field, the particles organized into a more tightly packed, ordered structure. Scientists could then measure and plot the column growth over time.

The space station's microgravity environment was critical to understanding the behavior of self-assembly in toggled fields. Microgravity slows down the movement of colloidal mixtures, allowing researchers to understand how they interact and how to control the tiny particles on the ground. You cannot do these experiments on Earth because the nanoparticles would settle out too quickly because of gravity.

At first, the particles in the fluid form long, thin chains. As the magnetic field is applied, the magnetic dipoles in the particles cause these singular chains to grow parallel to the applied field. The chains parallel to each other interact and bond together over time. These "bundles" of chains become more like columns when the magnetic field is toggled on and off. The columns grow in diameter with time as they are exposed to a pulsed, magnetic field. This self-directed "bundling" was not observed until the InSPACE-2 investigation, which ended in 2009.

When the InSPACE study began, it identified a pulsing phenomenon that had never been seen before. Work continued with InSPACE-2 and -3 investigations to further observe how magnetic fields impact colloidal, self-assembly phase transitions. By better understanding how these fluids "bundle" themselves into solid-like states in response to magnetic pulses, researchers have insight into phase separation. This may lead them to new nanomaterials from these tiny building blocks for use on Earth.

With new manufacturing models resulting from InSPACE -2 and -3 studies, these models could be used to improve or develop active mechanical systems such as new brake systems, seat suspensions, stress transducers, robotics, rovers, airplane landing gears and vibration damping systems. It also has promise to engineer new nanomaterials for thermal barriers, energy harvesting and color displays.

Thanks to the InSPACE series of investigations into tiny things, fundamental science could advance these systems and improve how we ride, drive, fly and live—in a big way.

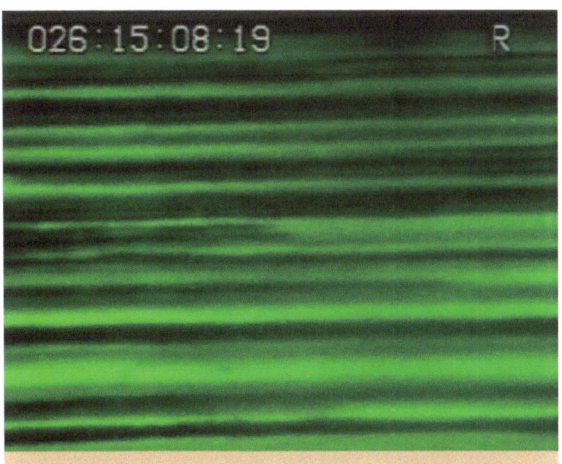

Investigating the Structure of Paramagnetic Aggregates from Colloidal Emulsions science video image of aggregates (columns forming). The black lines are the formed columns. The green background is from a green LED lamp used to provide lighting for the video camera.
Image credit: NASA

Satellites

The International Space Station (ISS) offers a unique platform for access to low-Earth orbit (LEO) through its Japanese Experiment Module (JEM) airlock working in coordination with the JEM robotic arm. This small airlock allows small devices such as CubeSats to be deployed into LEO while making the trip up to space in the relative comfort of a pressurized cargo container. This can have many benefits in reducing the cost to small satellite operators specifically in the number of launch testing and redundancy requirements for the developer. Lower cost leads to more financial incentive to enable small operators to design and prove out their technology in space.

Deploying small satellites from ISS

Traditional satellites require complex systems and often, the resources of a dedicated launch vehicle to find their way into orbit. However, with some help from the International Space Station, a new class of small satellites is changing the model for how we launch new technologies into space. CubeSats, small, less than 50 centimeter and mostly 10 centimeter (4 inch) cubic satellites, have an alternative way of being deployed. Some are deployed into orbit from the space station using a robotic arm. The satellites are transported to station in soft-sided bags by cargo ships such as Japan's H-II Transfer Vehicle (HTV). And at an appropriate time later, the satellites are taken out from the station's cabin, and the Japanese Experiment Module (JEM) Robotic Manipulator System (JEMRMS) aims the satellites at their planned orbits and releases them. The JEM Small Satellite Orbital Deployer, an ejecting system for small satellites, was developed by JAXA.

In the past, small satellites of a certain class have been launched by rockets as piggyback satellites. When a rocket can launch extra weight other than its main satellite, piggyback satellites are given their seats in

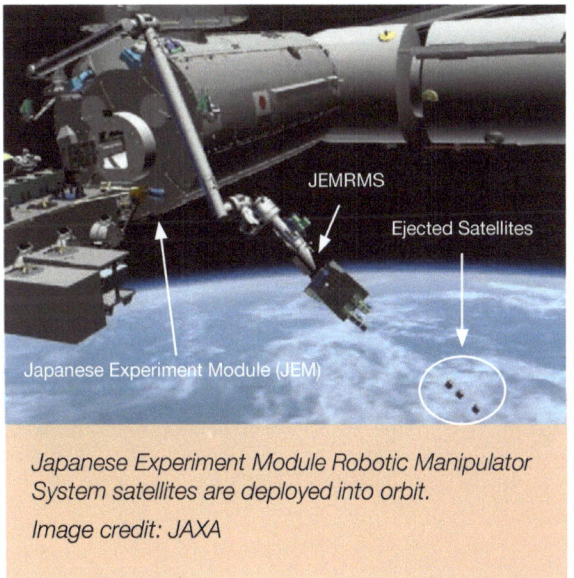

Japanese Experiment Module Robotic Manipulator System satellites are deployed into orbit.
Image credit: JAXA

the rocket and are thrown into the orbit after the main satellite is successfully deployed. The advantages of launching satellites from the space station by robotic arm compared to piggyback on a rocket include the option to choose the best timing of the small satellite's ejection without affecting the main satellite's timing. With limited space for small satellites to piggyback on rockets, the space station provides the additional benefit of having regularly scheduled cargo resupply flights on which the small satellites can more readily travel.

Any satellite must pass space environment tests to confirm that the satellite will survive the harsh environment during launch and its operational period in space. Among these, the vibration test that simulates

> Developers of small satellites have increased their use of space station deployers affording non-traditional users access to space.

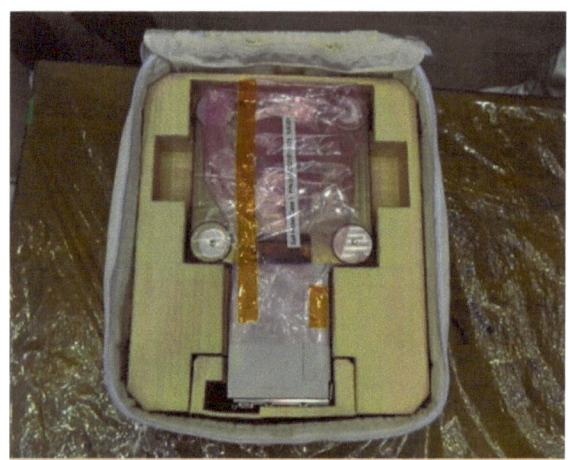

Satellite packed in a soft-sided bag.
Image credit: JAXA

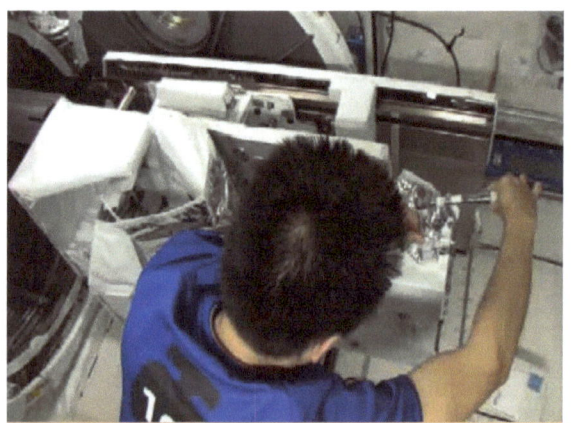

Crew in the International Space Station install a satellite onto the Japanese Experiment Module airlock table.
Image credit: JAXA

developers, such as college students, cannot afford to use expensive aerospace-rated electric parts to pass the vibration test.

An additional benefit to space station deployment of CubeSats is that after the CubeSats launch to space, astronauts aboard the orbiting outpost can perform quality checks on the hardware to ensure the small satellites are not damaged before deploying into space. One of the difficulties of developing free-flying satellites is that once they are launched, it is uncertain whether the satellite is still in good working order after enduring the launch vibration. With space station deployment, there is still opportunity to check out satellite systems and intervene before it deploys. This can allow designers to choose electric parts without the traditional space ratings, which can lower the total cost of development and expedite new space-qualified technology.

In order to launch a small satellite into orbit using the JEMRMS, first the space station cargo supply spacecraft, such as the HTV, delivers the satellite in a Cargo Transfer Bag. The satellites are stored in Cargo Transfer Bags in the space station cabin until time

FITSAT, one of the small satellites thrown into orbit by Japanese Experiment Module Robotic Manipulator System.
Image credit: Fukuoka Institute of Technology

vibrations experienced during launch, subjects the satellite to a rigorous level of agitation. Piggyback satellites are required to pass this test because they are installed in the same area as the main satellite. By contrast, satellites deployed from the space station are delivered by cargo spacecraft, where they are kept in a soft bag and buffered with packing material. The vibration level they experience during launch to the space station is less than that of piggyback satellites.

The relaxation of the vibration condition can be game-changing for small satellite developers because some for deployment. Following the final satellite checkout, the crew installs the small satellite into the JEM Small Satellite Orbital Deployer and places it on the JEM airlock table. Then, after the airlock is sealed, it opens to allow the airlock table to slide out of the cabin of the space station. The JEMRMS approaches the airlock table and grapples the satellite ejector. Next, the JEMRMS moves the satellite in the ejector into position for deployment into orbit. JEMRMS holds the specified attitude aiming at the satellite's orbit. Finally, ground operators send the command to the ejector to release the satellite.

In summary, introduction of this new method of satellite ejection using the JEM facility achieves the advantages of providing more frequent opportunities for small satellite deployment in low-Earth orbit, lowering the vibration test hurdles and providing the opportunity for a final checkout of the satellite before use.

As a result of the use of the space station, potential developers of small satellites have increased their use of the space station deployers, and universities, companies and other non-traditional space users are finding affordable access to space.

Pinpointing time and location

Lost phones and running late for appointments could become a thing of the past thanks to benefits originating from technology tested aboard the International Space Station. The Global Transmission Services 2 (GTS-2) experiment demonstrated that radio transmissions could be used to synchronize Earth-based clocks and watches and, eventually, to locate stolen cars and deactivate lost credit cards directly from space.

An antenna on the station currently transmits Coordinated Universal Time (UTC), also known as Greenwich Mean Time. These transmissions cover almost the entire Earth and can be received at a

Graphic of potential receivers for GTS signals from the International Space Station.
Image credit: Steinbeis-Transferzentrum Raumfahrt

particular location several times daily. The signal is strong enough to be received even by small wrist watches, and transmits accurate local time for different time zones, even taking into account daylight saving time. A unique code for each ground receiver verifies the authenticity of data and guarantees its secure transmission. Watches and clocks can automatically synchronize with these signals through receivers typically activated once per day or when the devices are turned on.

The GTS can function as theft protection by sending a message to a receiver chip in an electronic device such as a phone, car, or car key that shuts down the device. Even someone with an authentic key would not be able to steal the car to which it goes, because the signal makes the key unusable. That could reduce theft of car keys directly from drivers, which can sometimes turn violent.

The system also may be able to help deter theft of larger mobile items, such as shipping containers and truck trailers, by pinpointing their exact locations.

In addition to accurate setting of clocks and theft protection, other possible applications of the system include paging services, targeted broadcast of messages such as automobile recalls, remote control of various devices, container tracking and fleet-management services.

The system's ground receivers are capable of accurately determining the position of the space station based on its transmission of signals. This ability

> Global Transmission Services technologies can function as theft protection by shutting down stolen vehicles, making them unusable.

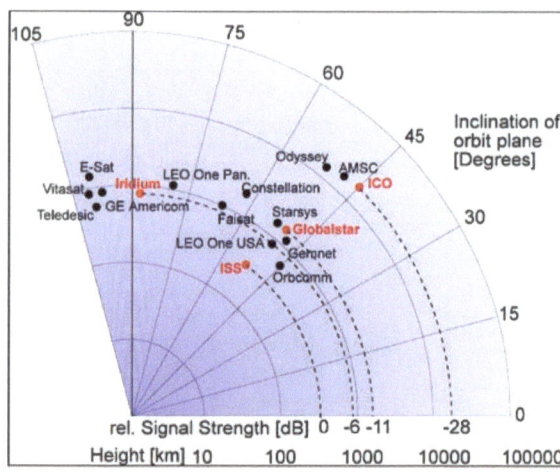

Orbit altitude and inclination of mobile radio satellite constellations and their relative signal strengths compared to that of the International Space Station.

Image credit: ESA

could be used in reverse to determine the location of a receiver from the station, a capability that one day might enable an orbiting spacecraft to navigate a ground vehicle on a planet below.

Using the space station for these global transmission applications offers several advantages over the use of other satellites. Because of the station's low orbit, every point between 70 degrees north and 70 degrees south latitude—most of the populated areas on Earth—can receive signals five to seven times a day. The transmissions also can be sent using very little power. Finally, because the station is manned, astronauts can exchange and maintain the device as needed.

The GTS experiment is supported by the company Fortis Swiss Watches, the German Aerospace Center (DLR), and the European Space Agency (ESA). Steinbeis Transfer-zentrum Raumfahrt provided experiment management, development and operation. The antenna unit was mounted on the Russian Service Module in December 1998, carried into space summer of 2000, and began operations in 2002, making this experiment one of the oldest aboard the station. Integration, launch and operation of GTS are now a cooperative undertaking with the Russian Institute of Applied Mathematics through the Russian Federal Space Agency (Roscosmos) and ESA's program of joint scientific investigations and experiments aboard the space station until 2020.

Space station technology demonstration could boost a new era of satellite-servicing

It may be called the Robotic Refueling Mission (RRM), but NASA's RRM was built to demonstrate much more than the clever ways space robots can refuel satellites. Following the success of this namesake task, RRM will demonstrate how space robots can replenish cryogen (a type of refrigerant) in the instruments of legacy satellites—existing, orbiting spacecraft not originally designed to be serviced.

> Remote refueling and robotics technology uses the space station as a test bed for technology research and development.

Delivery to Space Station and Installation

New hardware deliveries to the space station help to outfit the RRM module for the new set of operations. A new task board and the RRM On-Orbit Transfer Cage (ROTC), an original device designed to transfer hardware outside the space station, are added to the RRM module to produce the increased capabilities.

Astronauts mount the ROTC on the sliding table within the Japanese airlock and then install the task board onto the ROTC, giving the Canadian Dextre robot an easy platform from which to retrieve and subsequently install new hardware.

A second task board and a new device called the Visual Inspection Poseable Invertebrate Robot (VIPIR) are also added to the RRM module. This borescope inspection tool, built at the Satellite Servicing Capabilities Office (SSCO) at NASA's Goddard Space Flight Center in Greenbelt, Maryland, provides a set of eyes for internal satellite repair jobs. Both items are transferred and installed on RRM via the Japanese airlock, ROTC and Dextre.

The Robotic Refueling Mission investigation (center, on platform) uses Canadarm2, the International Space Station's robot arm, and the Canadian Dextre robot (right) to demonstrate satellite-servicing tasks.

Image credit: NASA

With the help of the twin-armed Dextre robot, the additional RRM task boards and the RRM tools, the RRM team works its way through intermediate steps leading up to cryogen replenishment. After retrofitting valves with new hardware, peering into dark places with the aid of VIPIR and creating a pressure-tight seal, the RRM and Dextre duo will stop short of actual cryogen transfer for this round of tasks.

RRM Phase 2 operations are scheduled to begin in 2014. Initial activities to demonstrate this in-orbit capability—cutting wires and removing caps—were completed in 2012 with the aid of the original RRM tools and activity boards.

Expanding Capabilities and Fleet Flexibility in Space

Cryogenic fluids are used on the ground and in space to make very sensitive cameras work better. However, in time, this extremely cold substance leaks out, and the camera no longer performs well. According to Benjamin Reed, deputy project manager of the SSCO, robotically replenishing these reserves would allow spacecraft instruments to last beyond their expiration date and ultimately permit satellites to perform longer.

Reed explains that both the government and commercial sectors are focused on expanding options for fleet operators. Operators can choose to extend the life of an aging observatory or spacecraft by use of a future cryogen-toting space tanker, instead of retiring or launching a new, costly one. The RRM demonstrations are an important step to eventually enabling that capability.

Preparing for a Servicing-Enabled Future

With the 2011 launch to the space station on the last space shuttle mission, RRM has been steadily practicing robotic satellite-servicing activities in orbit. RRM uses the space station as a test bed for technology research and development in a joint effort with the Canadian Space Agency.

NASA developed RRM to demonstrate how remotely operated robot mechanics could extend the lives of the hundreds of satellites residing in geosynchronous-Earth orbit (GEO). Costly assets traveling about 22,000 miles above Earth, GEO spacecraft deliver such essential services as weather reports, cell phone communications, television broadcasts, government communications and air traffic management. Servicing capabilities could greatly expand the options for government and commercial fleet operators in the future. They could potentially deliver satellite owners significant savings in spacecraft replacement and launch costs.

NASA continues to test capabilities for a new robotic servicing frontier. In conjunction with RRM, the SSCO team has been studying a conceptual servicing mission while building the necessary technologies, including an autonomous rendezvous and capture system, a propellant transfer system and specialized algorithms to orchestrate and synchronize satellite-servicing operations.

Transportation Technology

Combustion science is one of the longest running fields of research on the International Space Station (ISS). There is a long running campaign to understand just how both simple and more complex fuels burn in space. Understanding this process in microgravity helps us refine combustion models on Earth where gravity and turbulent buoyancy-driven convection flows make this process too difficult to model. Recent observations on ISS have shown that a phenomenon known as "cool flames" can be witnessed in the combustion chambers in orbit to understand how lower temperature burning could have significant applications towards more efficient fuel use and new combustion engine designs in the future.

Cool flame research aboard space station may lead to a cleaner environment on Earth

The anxious moments of trying to make the next service station, one eye on the fuel gauge, the low-fuel light staring at you, may become less frequent in the future. Even the choice of which fuel is better for the environment may be easier, thanks to droplet combustion research on the International Space Station; better mileage and a very real possibility of reduced pollution on Earth may be possible in the future.

> Thanks to the FLEX investigation in the reduced gravity environment of the space station, we have new insight into the mysteries of flames and fuel.

Researchers from academia working with NASA's Glenn Research Center in Cleveland, Ohio, conducted the Flame Extinguishing Experiments (FLEX and FLEX 2), which revealed some new insights into how fuel burns.

Led by principal investigator (PI) Forman Williams of the University of California, San Diego, who has studied combustion for more than 50 years, and co-PI Vedha Nayagam of Case Western Reserve University, Cleveland, Ohio, the results of the FLEX investigations revealed a never-before-seen, two-stage burning event. While a heptane droplet of fuel appeared to extinguish, it actually continued to burn without a visible flame. This knowledge could contribute to reduced pollution and better mileage in engine design because of improved prediction of flame behavior during combustion.

After decades of flame studies that have produced well-understood, theoretical models and numerical simulations, the FLEX flame investigations in microgravity produced this unexpected result. This is the first time scientists have observed large droplets (about three millimeters) of heptane fuel that had dual modes of combustion and extinction. The fire went out twice, once with and once without a visible flame. While the initial burn had a traditional hot flame, the second-stage vaporization was sustained by what is known as cool-flame, chemical heat release.

A cool flame is one that burns at about 600 degrees Celsius. To understand how cool this is, consider that a typical candle is about two-and-a-half times hotter, burning at around 1,400 degrees Celsius.

The phenomenon of the continued burning of heptane droplets after flame extinction in certain conditions was not anticipated when the study was designed. This result came during the FLEX investigation on the space station using the Multi-User Droplet Combustion Apparatus in the Combustion Integrated Rack (CIR). More recent FLEX experiments reveal similar two-stage burning phenomena with n-octane and decane fuels.

While burning the heptane droplets in the CIR, the first stage had a visible flame that eventually went out. Once the visible flame disappeared, the heptane droplet continued rapid quasi-steady vaporization without any visible flame. This ended abruptly at a point called second-stage extinction. At this point, a smaller droplet was left behind that either experienced normal, time-dependent evaporation or sometimes grew slightly through condensation of vapor in the cloud that formed upon extinction.

The new findings have been published and are available online in *Combustion and Flame*, the journal of the Combustion Institute. This new discovery will help scientists and engineers modify numerical models and better predict the behavior of flames, fuel and combustion. It also has many long-term implications both in space and on Earth. These findings can help with development of new technology to reduce pollution and increase gas mileage in internal combustion engines. Cool flame burning could also be used to partially oxidize the fuel for use in burners with reduced emissions and better control.

The Homogeneous Charge Compression Ignition (HCCI) engine combines diesel ignition with spark-ignition and can be used in any diesel engine, either stationary or for transportation. By merging these two technologies, engines could have the efficiency of burning diesel, while also providing reduced particulate and nitrogen oxide emissions. This could eliminate the need to burn diesel-fuel sprays, which are notorious for pollutant production, according to the researchers.

Thanks to the FLEX investigation in the reduced gravity environment of the space station, we have new insight into the mysteries of flames and fuel. Whether it's a candle, a campfire, or some other fuel source, the combustion process may be waiting for the right investigation to pry loose more secrets. Microgravity research may prove to be the tool that helps force those secrets free.

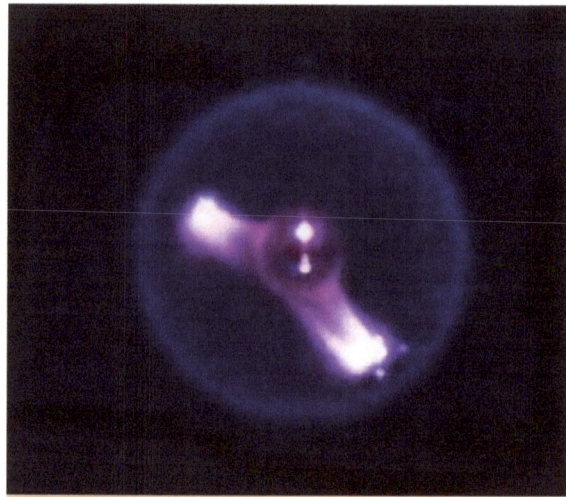

A burning heptane droplet during the Flame Extinguishing Experiments investigation on the International Space Station.
Image credit: NASA

Robotics

Key to enhancing human spaceflight missions is the ability of robots to work alongside the human crew to perform necessary tasks more efficiently. These tasks include those that are monotonous or risky and impose on the available time astronauts have to focus on science experiments. The International Space Station provides an excellent platform where these operational concepts and procedures can be developed, tested and evolved in an actual space environment while demonstrating robotic systems performance and reliability over the long duration. The precision and reliability requirements for space robotics led to dual-purpose technologies and advanced robotic capabilities for use on Earth.

Robonaut's potential shines in multiple space, medical and industrial applications

When scientists and engineers began developing Robonaut, a first humanoid robot for space exploration, they set out to create robotic capabilities for space exploration, but did not limit their design just for use in microgravity. Instead, they decided to lend a robotic hand, along with many other appendages and abilities, to those in need on Earth.

> Technologies developed for gloves, walking and telemedicine for Robonaut are being adapted for use on Earth.

The first Robonaut was a collaborative effort between NASA and the Defense Advanced Research Projects Agency. Though it was built for space exploration missions like performing skilled hand movements during Extravehicular Activity (EVA), or spacewalks, NASA has since gained significant expertise in expanding robotic technologies for space and Earth applications through successful creation of partnerships with outside organizations.

The latest iteration of Robonaut, Robonaut 2 (R2), was co-developed with General Motors (GM) through a Space Act Agreement. R2 is a faster, more dexterous robot, built for the microgravity environment to utilize human rated tools, assist with International Space Station activities and safely work side by side with astronauts. While R2 resides aboard the space station, many of the technologies developed for Robonaut and R2 are being adapted for use on Earth. Here are three examples:

Robo-Glove Technology

One of these, a robotic glove, or the RoboGlove, was developed as a grasp assist device after NASA and GM realized there was overlap between what astronauts needed in space and what factory workers could use on the ground. The RoboGlove can augment human tendons to help both astronauts and factory workers with grasping tasks and potentially minimize the risk of repetitive stress injuries.

Since astronauts wear pressurized spacesuit gloves during a spacewalk, they are exerting more force to hold a tool or tighten a screw, causing fatigue. The RoboGlove could help astronauts close their gloves and reduce the amount of effort they apply while conducting EVA tasks, much like the way power steering helps to steer a car.

At GM, factory workers on assembly lines are performing tasks like gripping tools repeatedly throughout their work day. These individuals are tiring more quickly during the day by either exerting a high amount of force at multiple intervals or exerting force for long periods of time. The RoboGlove may help the factory workers to grip a tool longer with less discomfort by reducing the amount of force that they need to exert. This could result in less fatigue and fewer stress injuries.

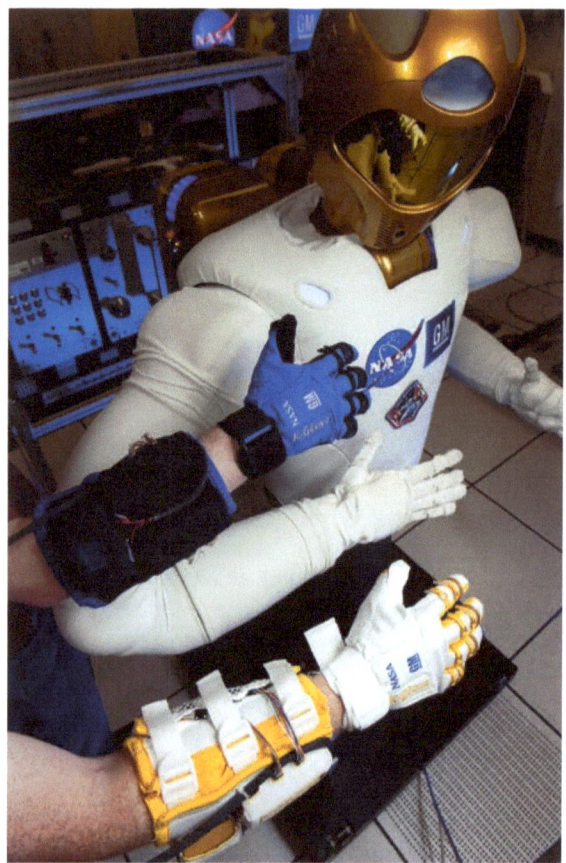

The Robo-Glove was built through the continuing partnership between NASA and General Motors. It uses R2 technology to decrease fatigue and stress when a human grasps an object.

Image credit: NASA

NASA and GM are working to find a supplier to make the patented RoboGlove. GM plans to use the glove technology in future advanced vehicle safety systems and manufacturing plant applications. NASA is experimenting with the technology in their Earth laboratory and integrating it into a working spacesuit glove for possible future use by crew members.

The RoboGlove also generates interest from the medical community. For instance, patients in rehabilitation may benefit from a device that helps them to recover their skills for grasping objects. Another potential application involves an adapted glove that could both open and close to help patients recovering from brain injury. NASA engineers have explored ways to adapt the glove for people with partial hand amputations, as well. A future partnership with a medical center or research institution could expand the technology of RoboGlove to medical settings, in addition to its use for space exploration and factory work at GM.

Robotic Exoskeleton

NASA and The Florida Institute for Human and Machine Cognition (IHMC), with the help of engineers from Oceaneering Space Systems of Houston, have jointly developed a robotic exoskeleton called X1. The X1 technology, derived from R2, may someday help astronauts stay healthier in space with the added benefit of assisting people with physical disabilities on Earth.

Currently in the research and development phase, X1 is a 57-pound robotic device that a human could wear over his or her body either to assist or inhibit movement in leg joints. Worn over the legs with a harness that extends up the back and around the shoulders, X1 has 10 degrees of freedom, or joints – four motorized joints at the hips and knees, and six passive joints that allow for sidestepping, turning and pointing, and flexing a foot.

Employing IHMC's experience in exoskeleton development for paraplegics, NASA and IHMC streamlined R2 arm technology. They made it slim enough to allow a person in a wheelchair to get out using the exoskeleton. The X1 device has the potential to produce enough force to allow for assisted walking over varied terrain to paraplegics or other patients in rehabilitation settings.

In addition to the IHMC and NASA applications of the X1 technology, researchers at the University of Houston (UH) are working to adapt an exoskeleton so it is controlled by brain signals. This type of exoskeleton would attach a device to a person's head and try to read signals that the brain sends to the legs to get them to move.

Telemedicine Applications

The Houston Methodist Research Institute (HMRI) and NASA worked together to adapt some of the technology used by R2 in space and on the ground. The research team tested R2 for use in telemedicine, conducting medical procedures through electronic communication by tasking R2 to perform an ultrasound scan of a medical mannequin and to use a syringe as part of a procedure.

NASA Project Engineer Shelley Rea demonstrates the X1 Robotic Exoskeleton, which could improve the mobility and strength of astronauts and paraplegics.
Image credit: NASA

 Watch these videos to learn more:

Robo-Glove Technology:
http://tinyurl.com/nasa-robo-glove

Exoskeleton Technology Applications:
http://tinyurl.com/nasa-exoskeleton-tech

Robonaut's Telemedicine Initiatives:
http://tinyurl.com/robonaut-telemedicine

With human control of the teleoperated R2, tasks were performed with accuracy and efficiency using R2's dexterity to apply the appropriate level of force and monitoring progress through R2's vision system. This demonstration of R2's capabilities could potentially allow physicians to conduct complex medical procedures on humans in remote locations on Earth or in space.

NASA's Space Technology Program is developing, testing and applying robotic technologies through innovative partnerships. They continue to look for new collaborative opportunities to leverage resources that will help all partners to increase their chances of making better products, as demonstrated by the numerous current applications of the R2 technology. Furthermore, using the International Space Station as a test bed for these robotic and future technologies will be vital to human exploration and beneficial to human health.

Student inspired by the ISS and the future of space exploration.
Image credit: NASA

Global Education

The International Space Station has a unique ability to capture the imaginations of both students and teachers worldwide. The presence of humans aboard the station provides a foundation for numerous educational activities aimed at capturing interest and motivating children towards the study of science, technology, engineering and mathematics (STEM). Projects such as the Amateur Radio on International Space Station (ARISS); Earth Knowledge Acquired by Middle School Students (EarthKAM); and Synchronized Position Hold, Engage, Reorient Experimental Satellites (SPHERES) Zero Robotics competition, among others, have allowed for global student, teacher and public access to space through student image acquisition and radio contacts with crew members. Projects like these and their accompanying educational materials are distributed to students around the world. Through the continued use of the station, we will challenge and inspire the next generation of scientists, engineers, writers, artists, politicians and explorers.

Inspiring the next generation of students with the International Space Station

In the 1960s, the innovators responsible for missions to the moon inspired homemade space helmets and backyard bottle rockets that sent toy soldiers a bit shy of low-Earth orbit. The innovations of the International Space Station (ISS), however, provide a more direct approach to opportunities for the next generation as they watch, learn and even participate in today's missions.

> ISS provides hands-on educational opportunities that encourage students to go beyond passive learning.

For more than a decade, the space station has provided hands-on educational opportunities that encourage students to go beyond passive learning, engaging them as interactive participants. In 2012, NASA published *Inspiring the Next Generation: International Space Station Education Opportunities and Accomplishments, 2000-2012*, which compiles these events to share with the public.

From 2000 through 2012, there have been more than 42 million students, 2.8 million teachers and 25,000 schools from 44 countries involved in education activities aboard the space station.

The publication was an international effort that details opportunities available today for students and summarizes those already completed. The comprehensive account includes education projects led by all the space station partners: NASA, the Canadian Space Agency, the European Space Agency, Japan Aerospace Exploration Agency and the Russian Federal Space Agency.

ISS Chief Scientist Julie Robinson engages students in International Space Station education activities.
Image credit: Sally Ride EarthKAM

Competitive student opportunities leverage real research to explain the depths of the scientific process; for instance, the Synchronized Position Hold, Engage, Reorient Experimental Satellites (SPHERES) ZeroRobotics challenge focuses on software programming skill development to guide robots through a virtual obstacle course aboard the space station. There are also contests that allow students to have their experiments performed in orbit, such as those proposed for Try ZeroG and the YouTube SpaceLab competitions, which were implemented on the space station in 2012. An inquiry-based approach to learning engages students and their communities, enabling them to contribute to the growing knowledge gained from research done aboard station.

These types of activities involve an investigative approach to learning and allows students to understand the true nature of science; gain in-depth knowledge of scientific concepts, laws, and theories; and develop interests, attitudes and "habits of mind" related to science and mathematics.

Other opportunities for station interaction include question-and-answer sessions via the Amateur Radio on the International Space Station (ARISS), which lets students contact astronauts on the station via ham radio. In-flight education downlink sessions through the NASA Education Office also enable student-crew communications, using live video feeds so communities can see the astronauts while speaking with them.

There are growing commercial opportunities, such as the Student Spaceflight Experiment Program in coordination with NanoRacks. This program provides opportunities for elementary and middle school students to propose and launch their own investigations to the space station.

The education publication is available in both hardcopy and through download via the International Space Station Research and Technology website at http://www.nasa.gov/iss-science.

Students from Charminade College Preparatory, West Hills, California, run preliminary variations of their experiment in the lab.

Image credit: Student Spaceflight Experiment Program

Inquiry-based Learning

From the launch of the first modules of ISS into orbit, students have been provided with a unique opportunity to get involved and participate in science and engineering projects. Many of these projects support inquiry-based learning—an approach to science education that allows students to ask questions, develop hypothesis-derived experiments, obtain supporting evidence, analyze data, and identify solutions or explanations. This approach to learning is well-published as one of the most effective ways to engage students to pursue careers in scientific and technology fields.

Student scientists receive unexpected results from research in space

YouTube is a great place to find bloopers, snuggly kitten or music videos. Now, it's also a place to post grand ideas for microgravity research studies. Two of those ideas actually got to fly on the International Space Station. The YouTube Space Lab competition provided such an opportunity, and three students saw their research performed aboard the orbiting laboratory.

> The YouTube Space Lab competition motivated thousands and selected three 14- to 15-year-old students to fly their microgravity research ideas on ISS.

Dorothy Chen, YouTube Space Lab winner, presents findings from her study on the anti-fungal properties of the bacteria Bacillus subtilis *in space at the 2013 International Space Station Research and Development Conference.*

Image credit: Center for the Advancement of Science in Space

For then-high school students Dorothy Chen and Sara Ma of Troy, Michigan, and Amr Mohamed of Alexandria, Egypt, the sky no longer is the limit for their research questions. Chen, Ma and Mohamed completed research investigations as winners of the YouTube Space Lab global contest. The competition invited 14- to 18-year-old students to submit two-minute videos via YouTube to propose physics or biology investigations for astronauts to perform aboard the space station. Their research was chosen out of more than 2,000 entries received from around the world.

The two winning studies looked at the anti-fungal properties of the bacteria *Bacillus subtilis (B. subtilis)* in microgravity, and the predatory behavior of *Salticus scenicus* and *Phiddipus johnsoni*, also known as a zebra jumping spider and a red-backed jumping spider, respectively. The results they received from their research informed them that microgravity can be a wily participant in a research study held in space aboard an orbiting laboratory.

Chen and Ma hypothesized that *B. subtilis*, a naturally occurring bacteria commonly used as an anti-fungal agent for agricultural crops, would have increased anti-fungal properties when grown in microgravity compared to the same bacteria produced on Earth.

Their testing also added phosphates and nitrates to the *B. subtilis* nutrient source to see if the additives affected growth and anti-fungal strength. The phosphates and nitrates acted as nutrients to potentially boost growth of the bacteria. The outcome of their investigation aboard the space station showed that the least amount of growth in the bacteria occurred in microgravity as compared to bacteria produced on Earth. They found that microgravity had no effect on the degree to which phosphates and nitrates affect *B. subtilis* growth.

Mohamed, now a college student in California, was excited about seeing his zebra jumping spider research in action while talking with Astronaut Sunita Williams in orbit.

Europe's alliance with space droids

Between video games and sci-fi movies, a group of young students find time for a robotic squadron of miniature satellites that come to life aboard the International Space Station, obeying their commands.

> High school students compete to program ISS droids and build critical engineering skills.

The European Space Agency (ESA) is participating in the NASA, Defense Advanced Research Projects Agency (DARPA), and Massachusetts Institute of Technology (MIT) Zero Robotics competition (http://www.nasa.gov/mission_pages/station/research/experiments/690.html). The competition is a chance for high school students to program droids for action on the space station. Synchronized Position Hold, Engage, Reorient Experimental Satellites (SPHEREs) are volleyball-sized satellites with their own power, propulsion, computers and navigation.

The challenge to remotely control them began in the United States, where an adventurous professor from MIT found inspiration in the Star Wars saga to create these intriguing robots. The mini-spacecraft have been used inside the station since 2006 to test autonomous rendezvous and docking maneuvers.

Now formation flying in zero gravity is a programming issue for European students, also. A number of schools from ESA member states create rival programs that control three SPHERES in real-time on the space station. A local SPHERES expert, familiar with the coding requirements for the droids, is assigned to each European school. Sponsored by ESA, several university staff members are being trained at MIT.

The competition is not only about feeding the satellites sets of commands; the local experts help students build critical engineering skills, such as problem solving, design thought process, operations training, and teamwork. Their results could lead to important

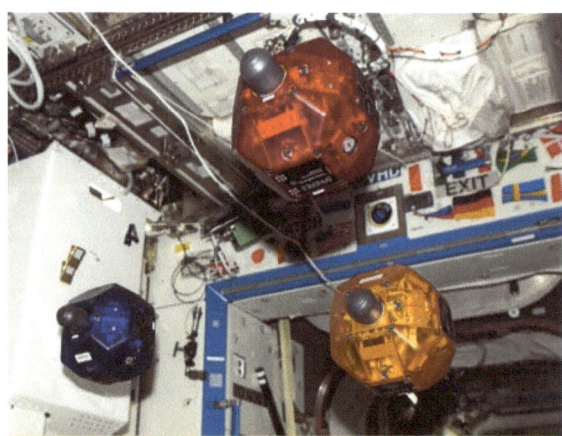

The three Synchronized Position Hold, Engage, Reorient Experimental Satellites (SPHEREs) on the space station are considered facilities. Two are used for Zero Robotics.
Image credit: NASA

Students participate in Synchronized Position Hold, Engage, Reorient Experimental Satellites (SPHERES) Zero Robotics.
Image credit: ESA

advances for satellite servicing and vehicle assembly in orbit.

Teams in the U.S. and in Europe test their algorithms under realistic microgravity conditions by competing in elimination rounds against each other with finals on both sides of the Atlantic.

The winners' software is uploaded and run in the three weightless SPHERES by astronauts on the station. The exciting final is streamed live at ESA's technology center in the Netherlands, European Space Research and Technology Center (ESTEC), and MIT.

Europe's alliance with the space droids is stronger than ever. The competition has given hundreds of high school students from across Europe and the United States the opportunity to operate droids in space by coding software. The success of the first rounds consolidates the station as a common scientific platform for students. They are embarking on a robotic future, and may the force be with them!

NASA has a HUNCH about student success in engineering

Several young science, technology, engineering and mathematics professionals entering the workforce right now are likely to have been motivated to enter those fields by the High school students United with NASA to Create Hardware (HUNCH) Program. HUNCH is a nationwide instructional partnership between NASA,

> Students build cost-effective hardware and soft goods for use on the ISS, launching careers and saving money for spaceflight programs.

European Space Agency astronaut Paolo Nespoli observes a can crusher, built by students in the High school students United with NASA to Create Hardware Program, during Expedition 26 aboard the International Space Station.
Image credit: NASA

Students from Cypress Ranch High School, Cypress, Texas, present mockup hardware to staff at NASA's Johnson Space Center built for NASA training programs as part of the High school students United with NASA to Create Hardware Program.

Image credit: NASA

high school and intermediate/middle school students to build cost-effective hardware and soft goods both for use on the International Space Station and for training of NASA astronauts and flight controllers.

With more than 11 years in existence, the popularity of HUNCH has grown to include 1,750 students in 77 schools across 24 different states. Trainees receive a hands-on opportunity that helps them strengthen their skills in science, technology, engineering and mathematics (STEM).

Students that participate in HUNCH learn to use and apply 3-D software, drafting, prototyping, welding, basic architecture, critical-thinking and problem-solving skills. NASA provides materials, equipment, mentoring and inspection oversight during the fabrication of these items. While students are building items for NASA, they are also building their self-confidence and interest as researchers.

To date, HUNCH participants have produced single-stowage lockers, cargo transfer bags, educational videos and experiments proposed to fly on the space station. Some standout projects include the design and fabrication of a disposable, collapsible glove box; an organizer for crew quarters on the space station; and a European Physiology Modules Rack trainer, which provides facilities for human physiology research. Since the beginning of HUNCH in 2003, hundreds of items for NASA have been produced.

In a culmination of skills learned as part of the HUNCH program, two students from Cypress Woods High School in Texas, Robert Lipham and Alie Derkowski, were selected to attend the Technology Student Association (TSA) National Competition in Orlando, Fla., to present the skills they acquired while building Microgravity Science Glove Box (MSG) trainers for NASA. HUNCH paired with the Center for the Advancement of Science in Space (CASIS) to provide funding for the students to showcase their engineering education endeavors.

Design updates made by the Lipham and Derkowski saved NASA money by streamlining the MSG trainers, which are mockups of space hardware for crew mission preparation. When the idea to create these items came to HUNCH, the cost estimate was $1 million for four MSG high-fidelity trainers. HUNCH provided NASA five MSG trainers for less than $250,000.

Every year, recognition ceremonies are held for all students and teachers that participate in the HUNCH program. The number of participants continues to grow annually, as well as the quality, quantity, and diversity of the products that students fabricate. While the recognition ceremonies recognize student work, they also acknowledge the educational benefits of NASA teaming up with students. This is often measured by the changes in the students' attitudes toward their own self-assurance and desire to enter STEM careers.

HUNCH is an innovative solution for inspiring the next generation of researchers and space explorers while providing money-savings and resource efficiencies for NASA. Schools can get involved through online application on the HUNCH website at http://www.nasahunch.com/.

Tomatosphere™: Sowing the seeds of discovery through student science

Home base on the moon. Boot prints on Mars. Visits to asteroids. With the world's space-faring nations looking beyond the International Space Station (ISS) to envision human missions to increasingly distant destinations, scientists have already begun to tackle the many challenges of sending humans farther and farther from our home planet. Missions to the ISS have made substantial contributions to our knowledge of how the human body adapts to microgravity for three, six or even 12 months, but taking steps further out into the solar system will require much longer expeditions. A human mission to Mars, for instance, will likely mean a six-month journey each way, coupled with a stay of about 18 months on the surface of the planet!

> The award-winning Tomatosphere™ educational project has helped students learn about science, space exploration, agriculture and nutrition.

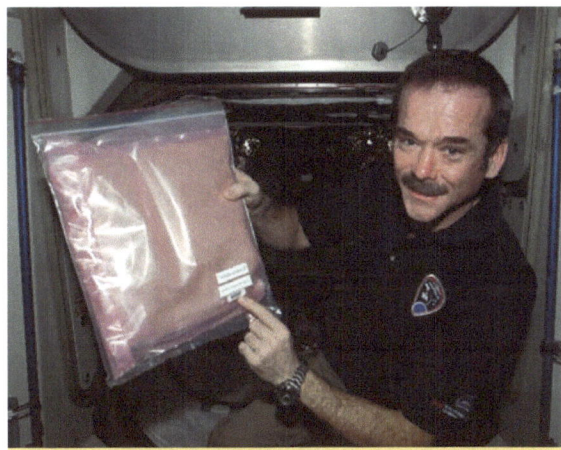

Chris Hadfield, Canadian astronaut and former commander of the International Space Station, poses with 600,000 tomato seeds for the Tomatosphere™ project, which returned to Earth with Hadfield in May 2013 after orbiting Earth for nine months aboard the station. The seeds will be grown by 16,000 classrooms in Canada and the U.S., a record number of participants for the project.
Image credit: CSA, NASA

Future crews on long-duration missions will need to be self-sufficient to stay safe and healthy. Since carrying two to three years' worth of food would be expensive and impractical, astronauts will have to grow their own food en route to their destination. Space farming may sound futuristic, but in the closed environment of a spacecraft, plants could make a huge contribution to life-support systems. Not only do plants provide food, water and oxygen, they also recycle carbon dioxide and waste. But how do you grow plants effectively in the radiation-filled environment of space? Which plants are best suited for space missions? What type of seeds would be able to withstand the journey and still germinate? What if we could recruit the next generation of astronauts, scientists and engineers to help solve the problem?

Since 2001, the award-winning Tomatosphere™ educational project has done just that. An estimated 3 million students in Canada and the United States have helped researchers gather data to address these questions, while learning about science, space exploration, agriculture and nutrition. Tomatosphere™ provides students with two sets of tomato seeds: one set that has been exposed to space or space-simulated environments as well as a control group for comparison. Without knowing which set is which, students grow the seedlings in their classrooms, measuring a variety of information about the tomato plants, the germination rates, growth patterns and vigor of the seeds. This methodology, known as a "blind study," allows the mystery of the project to be real science for the students. Each class submits their results to the project's website to be shared with scientists studying horticulture and environmental biology.

Grade 3 students measure their tomato plants as part of the Tomatosphere™ experiment.
Image credit: Tomatosphere™

The project's baseline experiment investigates the germination rate of the seeds; however, supporting materials have been developed to allow educators from grades 3 to 10 to build on student understanding of a variety of topics, from the science of plants to the science of nutrition to the science of ecosystems.

Tomatosphere™'s hands-on approach to learning gives students a taste for science and space research. In addition to being rewarded with their very own "space tomatoes" to bring home, the students participating in Tomatosphere™ today know that they have each made a personal contribution to assisting space exploration in the future. And perhaps one day, an astronaut biting into a fresh, juicy tomato on the surface of the Red Planet will thank them.

Tomatosphere™ is sponsored by the Heinz Canada, HeinzSeed, Stokes Seeds, the University of Guelph, Let's Talk Science, First the Seed Foundation and the Canadian Space Agency (CSA).

The above photo of The Bahamas is one of the pictures taken during the 2013 spring mission.
Image credit: Sally Ride EarthKAM

Students photograph Earth from space via Sally Ride EarthKAM program

Sally Ride EarthKAM (Earth Knowledge Acquired by Middle School Students) is a NASA educational outreach program enabling students, teachers and the public to become researchers, focused on learning about Earth from the unique perspective of space. During the Sally Ride EarthKAM missions, middle school students around the world request images of specific locations on Earth. The entire collection of Sally Ride EarthKAM images is available in a searchable image archive. This image collection and accompanying learning guides and activities are extraordinary resources to engage students in Earth and space science, geography, social studies, mathematics, communications, and art. Four missions are offered per year.

Sally Ride EarthKAM uses a Nikon D2Xs digital camera mounted in the Window Observational Research Facility (WORF), which uses the science window located in the U.S. Destiny Laboratory. This window's high-quality optics capabilities allow the camera to take high-resolution photographs of the Earth using commands sent from the students via the online program. Students and educators then use the photos as supplements to standard course materials, offering them an opportunity to participate in space missions and various investigative projects. Creators of Sally Ride EarthKAM hope that combining the excitement of this space station experience with middle-school education will inspire a new set of explorers, scientists and engineers.

Students use Sally Ride EarthKAM to learn about spacecraft orbits and Earth photography through the active use of Web-based tools and resources. With the help of their teachers, they identify a target location and then must track the orbit of the station, reference maps and atlases, and check the weather prior to making their image request. These requests funnel to another set of students, this time at the University of California at San Diego. These college students run the Sally Ride EarthKAM Mission Operations Center for the project. Here they compile the requests into a camera control file and, with the help of NASA's Johnson Space Center, uplink the requests to a computer aboard the space station.

Requests ultimately transmit to the digital camera, which then takes the desired images and transfers them back to the station computer for downlink to Sally Ride EarthKAM computers on the ground. This entire relay process is usually complete within a few

> Middle school students track the orbit of the station, reference maps and atlases, and check the weather before making an imagery request.

Two students from Good Shepherd School in Alberta, Canada, participate in Sally Ride EarthKAM.
Image credit: NASA

> Students from countries across Asia compete to see their microgravity experiments conducted on ISS.

hours and the photos are available online for both the participating schools and for the public to enjoy.

Sally Ride EarthKAM was initiated in 1995 and originally called KidSat. The KidSat camera flew on three space shuttle flights (STS 76, 81 and 86) to test its feasibility and was later moved to the International Space Station (ISS) and renamed ISS EarthKAM. In 2013, the program was renamed once more, this time to Sally Ride EarthKAM to honor the late Dr. Sally K. Ride, America's first woman in space and the program's creator. Dr. Ride passed away July 23, 2012.

Since its first space station expedition in March 2001 to the present, Sally Ride EarthKAM has touched the lives of nearly 300,000 student participants and an unknown number of online followers. The program also has a strong international presence with users from 74 countries to date. Interested viewers of Sally Ride EarthKAM images and educators interested in participating have the opportunity to register online at https://earthkam.ucsd.edu/.

Try zero G 2: Igniting the passion of the next generation in Asia

Twelve-year-old Lily Thornton sat with her mother in an Australian airport getting ready to leave her homeland for the first time while waiting to board a plane to Japan. Such a young girl could be quite nervous about going to a completely different environment or the ensuing language barrier that she would face once she disembarked the plane, but Lily had on her mind something that most 12-year-old girls rarely have to think about. Lily was going to see her microgravity experiment conducted on the International Space Station (ISS) through a program called Try Zero Gravity (Try Zero G).

Try Zero G is a program aimed at children and educators to learn about ISS and its educational experiments. In 2009, Japan Aerospace Exploration Agency (JAXA) started a domestic, micro-gravity program, Try Zero G, using the Japanese Experiment Module, Kibo. It is a brand new, innovative, educational program that gives the public the opportunity to participate in ISS experiments in Japan. This program has successfully stimulated public interest. A total of 1,597 ideas were submitted with the hope of being implemented in 2009. Among those, 16 were selected, and the domestic Try Zero G idea was conducted by a Japanese astronaut and continued thereafter.

With the belief that educational activities conducted on ISS are a benefit for humanity, JAXA opened up

Lily Thornton, age 12, of Australia at Tsukuba Space Center, Japan Aeropsace Exploration Agency (JAXA).
Image credit: JAXA

the Try Zero G program to Asian nations, implemented from 2011 under the framework of the Asia-Pacific Regional Space Agency Forum (APRSAF). The first Try Zero G for Asian nations was announced in APRSAF-17 in 2010 and 10 ideas from three countries were submitted. Five ideas were conducted by JAXA's astronaut Satoshi Furukawa and implemented on Sept. 22, 2011. The second Try Zero G was announced in APRSAF-18 in Singapore right after the first Try Zero G was conducted. The number of applications increased, and JAXA received 127 applications from four countries. Eight ideas were implemented by JAXA's astronaut Akihiko Hoshide on Nov. 15, 2012. Every Try Zero G, JAXA welcomes those whose ideas have been selected to observe their experiments at the JAXA facility in Tsukuba Space Center, Japan.

Lily's idea, "Weight Station," observes the behavior of a spring balance under microgravity. Having her idea selected for the ISS experiment became big news in Lily's small town and in her school in Victoria, Australia. Once she heard of the opportunity to watch the downlink in JAXA's facility, she wished to see her experiment performed with her own eyes. Her school and the whole town held fundraisers for her to earn her travel expense to go to Japan to watch her dream come true.

On Nov. 27, 2012, Lily and her mother, Elise Thornton, were able to travel to Japan to visit JAXA's Tsukuba Space Center. Together they toured the space center and watched the downlinked video of Lily's experiment conducted by astronaut Akihiko Hoshide. During the tour, they had the opportunity to talk to the flight director of the mission control room in Tsukuba Space Center. The flight director talked about ISS and of microgravity conditions. Lily and her mother listened in rapt attention and had a chance to ask questions further expanding their knowledge of ISS and microgravity conditions. Lily found the tour enjoyable and interesting. Elise said that Lily dreams of one day becoming a space robotics engineer. We hope Lily stays true to her dreams and realizes her efforts!

Inspiration

Conducting education activities is not the reason the space station was built, but the presence of astronauts aboard the ISS serves as an inspiration to students and their teachers worldwide. Having the opportunity to connect with crew members real-time, either through "live" downlinks or simply speaking via a ham radio, ignites the imagination of students about space exploration and its application to the fields of science, technology and engineering.

Asian students work with astronauts in space missions

Methawi Chomthong of Mahidol Wittayanusorn School in Thailand plants chili seeds to observe how they grow while Leonita Swandjaja of Bandung Institute of Technology in Indonesia distributes tomato seeds to primary school pupils. These are seeds that have traveled in space and many students and pupils in the Asia-Pacific region have enthusiastically planted and nurtured these "space seeds." The Space Seeds for Asian Future (SSAF) (http://iss.jaxa.jp/en/kuoa/ssaf/) is a joint program run by space agencies and institutions for science education in the region.

They observed the growth of their plants to see if there was any difference between the ground and space seedlings. Karen showed the seedlings in the box on a video camera. She then pulled out some seedlings and examined how strong their stems were. The video image of the operation was downlinked to the Tsukuba Space Center, Japan Aerospace Exploration Agency, Japan. The ground staff observed the space seedlings as conveniently as if they were side by side with her. These downlinked video images were distributed to the organizations participating in the SSAF2013 program, and a timeline was set for showing the video to students, making them feel like they were working with an astronaut.

> Through collaboration with astronauts, students compare how seedlings in space look different from those they grew on the ground.

SSAF does more than simply sending seeds into space and bringing them back to Earth. It requires collaboration between astronauts and students on the ground. In September 2013, astronaut Dr. Karen Nyberg pulled out a box from a stowage rack in the Japanese Experiment Module, Kibo, of the International Space Station. The box contained seedlings of Azuki, small red beans, that grew seven days after being watered and kept under dark conditions. In parallel on the ground, students prepared their own plant boxes and started cultivating their own seeds.

Students from Osaka City University, Osaka, Japan, monitor the SSAF2013 experiment at Tsukuba Space Center as members of the ground crew, who have played an important role in developing the experiment protocol and preparing the plant materials.
Image credit: JAXA

In Malaysia, the National Space Agency (ANGKASA) held a competition to help students develop their skills in science research with SSAF2013. A total of 79 teams, each consisting of five members from 25 primary schools and 54 secondary schools, participated in the competition. In other countries, including Australia, Indonesia, Japan, New Zealand, Philippines, Thailand and Vietnam, various age groups of students and pupils learned the scientific method through this experience. These young people play an important role, not only in space technology, but also in other fields of science and technology for the future development of those areas.

Malaysian students set their experiment.
Image credit: MARA Junior Science College, Royal Malaysian Police, Kulim, Malaysia

The results showed that the seedlings in space looked quite different from those on the ground. Students understood the wonderful capability of such tiny seeds by witnessing that they were able to adjust to various gravitational conditions. Although stricter control of experimental conditions is required for the more involved science, the observations in SSAF2013 still offer many hints to scientists who are developing their new research projects.

Educational benefits of the space experiment 'Shadow-beacon' on ISS

The use of spaceflight for stimulation of public interest to advance science and education is a common practice among the global space agencies.

> ISS provides opportunities to stimulate student interest and participation in science, technology, engineering, and math (STEM).

Particularly at the Russian Segment of the International space station (ISS RS) the space experiment Shadow-beacon (SE) has been performing in series since 2011 for the scientific and educational purposes.

As an on-board radio beacon transmits VHF sounding signals of a 145-megahertz range, ground participants can register moments of signal appearance, follow the signal until it vanishes using the time marks, and send this information, along with data on its geographical position, to the Information Storing Center on Earth. Every operating sequence would take up to 20 minutes, while ISS is passing over the given continental measuring field.

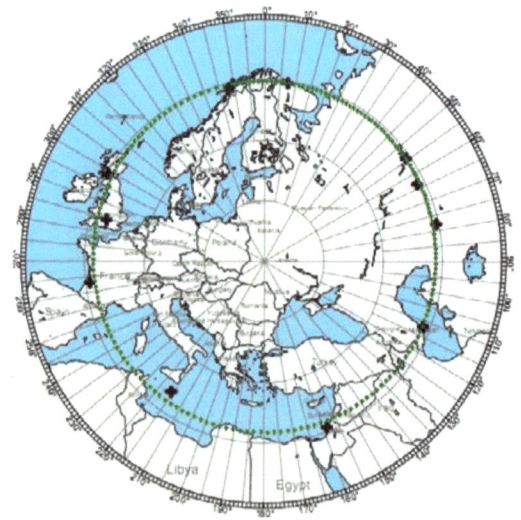

A typical result of construction of International Space Station experimental footprint contour on the Earth surface in the azimuthal projection. European measuring field, 27.11.11. Current moment 05.02.23 UTC. Position of undersatellite point: Ukraine, latitude 50.45, longitude 26.54.
Image credit: FGUP TsNIIMash

Onboard Amateur radio equipment "Sputnik."
Image credit: FGUP TsNIIMash

The data provides instant mutual position of the ISS, and each receiver allows definition of experimental borders of the station's footprint, i.e., "illuminated" spot on the Earth's surface. With many ground receivers, Shadow-beacon simulates a "multibeam" method of radio sounding of undersatellite space. Its basic properties are evaluated by comparing obtained experimental and calculated contours of the ISS footprint. Shadow-beacon is a developing methodology for the future experiment "Shadow," which will use radio waves scattering in an artificial plasma.

Possible application of this radio-sounding method is observation of interference in radio communication caused by plasma plumes of perspective electric thrusters, which are planned to be used for Martian expeditions. Exclusive simplicity of the radio sounding method allows the opportunity to carry out Shadow-beacon by non-professional operators (radio amateurs) and includes participation by educational programs. Therefore, the goals and objectives of the SE are both scientific and educational in nature.

Observations gained in Shadow-beacon sessions using the in-orbit "Sputnik" hardware between 2011-2013 involved around 70 ground operators in the testing and development of special software for construction of experimental ISS footprint contours on the Earth's surface. In a November 2011 series of Shadow-beacon sessions, laboratory curriculum for students was tested and students demonstrated the Shadow-beacon procedure as an extracurricular activity.

For more information about the formulation and conditions of the experiment, the sessions schedule, registration instructions, information on the progress of its implementation, or for training materials, visit the website at http://knts.tsniimash.ru/shadow/en/Default.aspx.

To date, the Shadow-beacon website has received over 160 applications for participation from private and club amateur radio stations. This includes educational institutions interested in using Shadow-beacon procedures in the learning process. In order to improve the methodology for educational purposes and to strengthen its social significance, the developers hold classes in which students are directly involved in the process of data registration and analysis, and in preparation and sending resulting reports to amateur radio operators. These operators are registered participants of the experiment who will be invited to help the neighboring schools conduct space lessons.

Expanding the ISS educational laboratory to orbital heights through use of programs like Shadow-beacon provides opportunities to stimulate student interest and participation in the educational process.

Students from Moscow's Center of Social Aid to family and Children "Pechatniki" take part in registration of sounding signals from orbit by field station ra3awc in the November 2011 series of Shadow-beacon (SE).
Image credit: FGUP TsNIIMash

Students get fit the astronaut way

When you think of NASA, you likely picture either the space shuttle or the International Space Station. Perhaps you have images of planets and galaxies flashing before your mind's eye. NASA's Mission X: Train Like an Astronaut (TLA), however, focuses a little closer to home. Working with the schools in local neighborhoods and around the world, Mission X applies the same skills used to train astronauts, as well as the excitement of space exploration, to motivate over 15,000 students in 140 cities around the globe to live a healthier lifestyle.

> Students ages 8 through 12 learn about the importance of hydration, bone health, balanced nutrition, and fitness.

Participants at The Resource Center (TRC) in Jamestown, New York, celebrate their kick-off of Mission X 2013.
Image credit: TRC

Led by NASA's Human Research Program, the TLA project includes physical education activities and educational modules on an interactive website (www.trainlikeanastronaut.org). The activities and modules are available in 15 different languages for participants in 22 countries including, the United States, Austria, Belgium, Colombia, the Czech Republic, Denmark, France, Germany, Indonesia, Ireland, Italy, Japan, Kazakhstan, the Netherlands, Norway, Portugal, Puerto Rico, Russia, Spain, Sweden, Switzerland, and the United Kingdom. This growing list of countries brings the TLA program closer to its goal of making kinesiology and nutrition fun for children by encouraging them to train like astronauts!

Chuck Lloyd, the NASA program manager responsible for the project, commented on how the space program excites students, prompting their active participation. Mission X is all about inspiring and educating our youth about living a healthy lifestyle with a focus on improving their overall daily physical activity with the Mission X physical activities.

Students ranging from 8 to 12 years old learn about the science behind their activities, including the importance of hydration, bone health and balanced nutrition. The activities have also been adapted for individuals with unique needs. Known as "fit explorers," the participants stay motivated with fun ways to gauge their success. For instance, they can see what other schools are doing on the Train Like An Astronaut blog. Fit explorers log their accumulated activity points over the course of the program to help an online cartoon astronaut, known as Astro Charlie, walk to the moon. Astro Charlie has made it to the moon every year—a distance of 238,857 miles (384,403 km) or 478 million steps—and he's still going!

Fit explorers learn that astronauts train before, during and after missions to maintain top physical health via good nutrition, rest and physical activity to function

Third- and fourth-grade students in Japan participate in the "Jump for the Moon" activity.
Image credit: JAXA

in the demanding environment of microgravity. Lloyd makes the connection of such health-centric mindsets for everyone, even those not planning to launch into space. Program success is met when our youth makes smart choices by balancing the amount of work, play and sleep they get to maintain peak performance. Education is critical to our youth and to our communities to ensure we have tomorrow's workforce and technical leadership poised to address the rigors of our societies.

Mission X was piloted in 2011. The challenge takes place from January to March when the participants from around the world complete the activities and team up to help Astro Charlie arrive at the moon. Each year Mission X continues to grow in the number of participating children and adults, countries and languages.

During the 2014 Challenge, participants were able to follow NASA astronaut, Mike Hopkins, who encouraged Fit Explorers to join him in a lifelong journey to improve health and fitness. In preparation for his mission to the ISS, a series of videos and products were produced to showcase his astronaut training. While on the station, he connected with participants to share his experiences in space. After his return to Earth, you can also learn what it took for him to return his strength and fitness to pre-mission levels. We encourage you to follow the TLA Facebook and Twitter pages at http://www.facebook.com/trainastronaut and https://twitter.com/trainastronaut.

Inspiring youth with a call to the International Space Station

Ever since the Amateur Radio on the International Space Station (ARISS) hardware was first launched aboard Space Shuttle *Atlantis* on STS-106 and transferred to the space station for use by its first crew, it has been used regularly to perform contacts with education organizations. The overall goal of ARISS contacts is to get students interested in science, technology, engineering, and mathematics (STEM) by allowing them to talk directly with the crews living and working aboard the station.

The ARISS conversations usually last for about 10 minutes. During that time, crew aboard the station answer students' questions as an audience of students and community members look on.

An ARISS contact takes place as a part of a comprehensive suite of education activities. In preparation for these exchanges, students learn about the space station and the research conducted aboard the space station. In addition to learning about life in space, the students learn about radio waves and how amateur radio works. The ARISS program is all about inspiring and encouraging by reaching the community and providing a chance for schools to interact with local technical experts. It also brings the space program to their front door.

In order for ARISS to work, the station must pass over the Earth-bound communicators during amateur radio transmissions to relay signals between the station's ham radio and ground receivers. Other factors, including weather, crew availability, and the schedules of vehicles visiting the space station, drive the timing of the scheduled transmissions. During this pass, an average of 18 questions can be answered, depending

> Students get interested in science, technology, engineering and mathematics by talking directly with crews living and working on the space station.

Astronaut Sunita L. Williams, flight engineer for Expeditions 14 and 15, talks with students at the International School of Brussels in Belgium during an Amateur Radio on the International Space Station session in the Zvezda Service Module.
Image credit: NASA

A student talks to a crew member aboard the International Space Station during an Amateur Radio on the International Space Station (ARISS) contact.
Image credit: ARISS

Calling cosmonauts from home

Educating future generations of scientists, technologists, engineers and mathematicians is a global effort—one that includes the contributions of the Russian Federal Space Agency, or Roscosmos. One of the main objectives of activities aboard the International Space Station is the implementation of education and outreach projects that contribute to attracting young people to studying science. These projects also help create modern, high-technology equipment and increase society's support of space programs in general and in the space station program, particularly. Currently aboard the Russian segment of the station are four space investigations that have educational components: Coulomb Crystal, Shadow-beacon, MAI-75 and Great Start. These experiments continue to demonstrate great benefits in capturing the imagination of students across Russia.

> Future generations of scientists, technologists, engineers and mathematicians get their start through global communication.

on the complexity of the query. To date, the space station has held more than 800 ARISS sessions with students in over 40 countries around the world.

The downlink audio from ARISS contacts can be heard by anyone in range with basic receiving equipment, transmissions broadcast on 145.800 megahertz. Interested parties can also catch a broadcast via EchoLink and the Internet Radio Linking Project (IRLP) amateur radio networks or on the Internet, when available.

For students who have never thought about space exploration, being involved in an amateur radio event such as this can be an eye-opener and pave the way for them to dare to dream and for those dreams to come true.

U.S. educators interested in participating in an ARISS communication can contact NASA's Teaching From Space Office for a proposal packet. International schools should submit applications via the ARISS website for consideration. Submissions are due in July and January of each year.

Coulomb Crystal is an investigation aimed at studying the dynamics of solid, dispersed environments in an inhomogeneous, magnetic field in microgravity. Pilot studies aboard the station explore the structural properties of Coulomb clusters—liquid crystal phase transitions, wave processes and the physical and mechanical characteristics of its heating mechanism, to just name a few. Students at all levels, including secondary school and college, have had the opportunity to prepare and conduct the experiment on the ground.

Shadow-beacon is a VHF radio beacon that allows amateur radio enthusiasts to communicate with crew members aboard the station. The presence of this equipment on the Russian segment of the station serves as a learning tool for students in the area of space communications. Students learn about the conditions of the admission-transfer of the radio beacon using the world amateur radio network. They also study the characteristics and spatial distribution of the intensity of the radio broadcast and rebroadcast from the onboard transceiver.

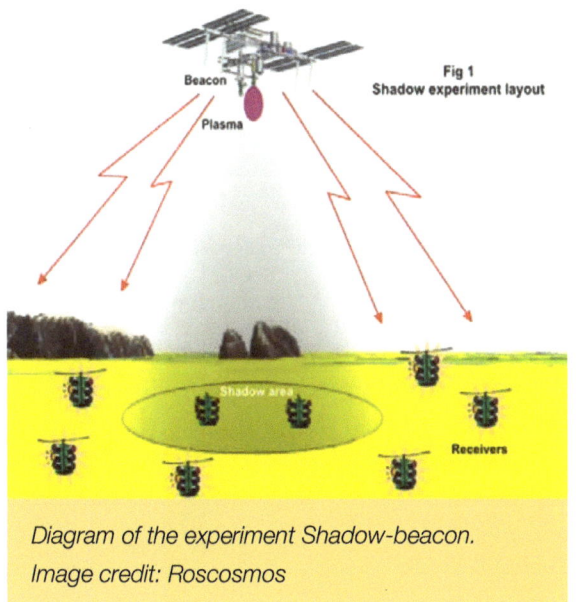

Diagram of the experiment Shadow-beacon.
Image credit: Roscosmos

Image taken by the reception and processing center MAI.
Image credit: Roscosmos

Along the lines of Shadow-beacon, MAI-75 is also part of the suite of communication equipment housed aboard the Russian segment of the station. This investigation allows for a system of quick video downlinks from space in near-real-time. This network distribution affords students and amateur radio operators from all over Russia the opportunity to learn first-hand from space explorers what it is like to live and work in space. The use of Earth images from space are also an effective source of inspiration and motivation for students.

Great Start is an investigation aimed at popularizing the achievements of cosmonautics in Russia and in the world. Developed with the preparation of a special questionnaire, this experiment allows the general public an opportunity to express its opinion regarding the first human flight in space, a great event in human history. The public also gets acquainted with the results of scientific investigations conducted aboard the space station. Great Start promotes and enhances international cooperation on the space station for further integration of Russia into the world of cultural, educational and scientific relations. As a result, there will be scientific and educational workshops held to popularize the achievements of Russian human spaceflight with the involvement of the general population, including students and specialists in various areas of possible utilization of the results of space missions.

There are several more experiments planned within the framework of the educational program: ecology-education, which includes student experiments that demonstrate the conduct of airborne, microscopic particle suspensions in microgravity; chemistry-education, which includes student experiments that capture microgravity structural elements in the specified form on the basis of polymer composite materials and diffusion; and diffusion, which is an educational demonstration of the process of diffusion in liquid environments in weightlessness. These educational projects involve hundreds, if not thousands, of students from all regions of Russia. Like all of our space station global, regional and national education programs, these experiments serve to inspire and motivate students to pursue careers in science, technology, engineering and mathematics (STEM).

MAI-75 experiment, main results and prospects for development in education

The MAI-75 experiment develops and validates the concepts for designing and operating an innovative telecommunication satellite system at the Moscow Aviation Institute (MAI) to support video information broadcasting from space in real-time to a wide range of users within Russia's academic mobile communications and internet user communities

The MAI-75 space experiment ("Spacecraft and Modern Personal Communication Technologies") has been carried out on the International Space Station Russian Segment (ISS RS) since 2005.

> Students were able to immerse themselves in real-life science and engineering applications while learning management and leadership skills.

Russian Cosmonaut M. V. Tiurin during a communication session with the Moscow Aviation Institute.

Image credit: Moscow Aviation Institute (National Research University)

The MAI-75 experiment is carried out using a notebook computer on the ISS RS, which stores and prepares the photos and videos that are then transmitted to Earth using the ham radio communication system, the primary component of which is the onboard Kenwood TM D700 transceiver of the "Sputnik" ham radio system within the 144-146/430-440 MHz bands.

The experiment used a communication channel operated on the ham radio frequencies, allowing to significantly expand the number of experiment participants both in Russia and globally. The experiment results can be obtained at a work station. All that is needed is transceiver equipment that operates on VHF ham radio frequencies.

During the experiment, a total of 120 communication sessions were carried out between the ISS RS and the MAI Data Reception and Processing Center, each with a duration of 9-15 minutes, and over 240 video images were received ranging in size from 14 to 94 KB.

After pre-processing, the images obtained are posted on a special website where they can be viewed, reviewed and processed by both educational program participants and common users. Besides the MAI Data Reception and Processing Center, the following were involved in the imagery reception process: Reception centers at higher education institutions in Moscow (M. V. Lomonosov Moscow State University, N. E. Bauman Moscow State Technology University); Krasnoyarsk (Siberian State Aerospace University); and Kursk (Kursk State Technology University); Reception centers at the Aerospace Technology Research Laboratory (Kaluga) of the Russian Defense Sports-Technology Organization; Gagarin Research and Test Cosmonaut Training Center (Star City); and S.P. Korolev RSC Energia (Korolev); Ham radio reception stations in Russia, Western Europe, Central America, and South-East Asia.

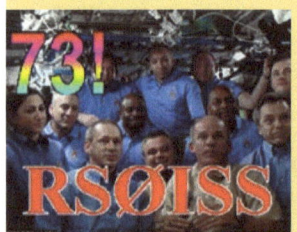   

Samples of images taken from the ISS RS by amateur radio communication channel during the sessions of SE "MAI-75."

Image credit: Moscow Aviation Institute (National Research University)

Some of the video images received are posted on the following site: http://www.issfanclub.com/image/tid/54.

During ground-based preparations for MAI-75 experiment sessions, the MAI teachers and students in conjunction with experts from RSC Energia and FGUP TsNIIMash led the development and testing of in-orbit procedures and cosmonaut training; development of software and hardware packages for use on Earth at global test sites and in space; and the development and verification of procedures and tests between the ground-based remote user terminals.

As a result of the first phase of the MAI-75 experiment, students were able to immerse themselves in real-life science and engineering applications, while learning management and leadership skills unique to the space station vehicle.

Space experiments such as these enable the secondary and higher-education systems to enhance the effectiveness of teaching natural sciences and to promote the interest of the public in the space programs implementation. The capabilities of modern information and communication technologies, particularly the Internet, and of the mobile (cellular) communications operators enable education program participants to work directly with the general-purpose video equipment deployed on the ISS. Using a Web interface and a special site, the program participants are able to control a digital camera installed on the ISS RS, based on both the Web-posted camera operation schedules and the ISS sub-satellite point movement data.

Centers of data reception from the International Space Station Russian Segment during MAI-75 experiment sessions.
Image credit: Moscow Aviation Institute (National Research University)

International Space Station:
Fostering commerce in space.

Economic Development of Space

While the International Space Station (ISS) has proven its value as a platform for a broad waterfront of research disciplines as well as technology development, it also provides an ideal opportunity to test new business relationships. This allows an opportunity to shift from a paradigm of government-funded, contractor-provided goods and services to a commercially provided, government-as-a-customer approach.

This interest in promoting a more commercially oriented market in low-Earth orbit (LEO) is driven by several goals. First, it can stimulate entirely new markets not achievable in the past. Second, it creates new stakeholders in spaceflight and represents great economic opportunity. Third, it ensures strong industrial capability not only for future spaceflight but also for the many related industries. Finally, and perhaps most importantly, it allows cross-pollination of ideas, processes, and best practices, as a foundation for economic development.

From commercial firms spending some of their research and development funds to conduct research on the space station, to commercial service providers selling unique services to users of the orbiting lab, the beginnings of a new economy in LEO are starting to emerge.

Commercial Service Providers

Evolution of the space station as a laboratory in the vanguard of research in microgravity relies on a new and growing number of commercial service providers. Rather than follow the traditional model of government-funded, contractor-provided hardware or capability, a number of firms are entering a new phase of development of LEO—establishing a market. In this model, commercial firms develop capabilities that are then offered to government users and also marketed widely to potential new users of the ISS as a research platform. The space station gains important new (or updated) capability, while the service provider gains a new market in which to offer its services.

Water production in space: Thirsting for a solution

Developing and maintaining water production on the International Space Station is vital for keeping the crew alive as well as supporting hygiene and equipment functions, yet it presents a bit of a challenge. Historically, about half of station's requirements were met by recycling used water and the rest by deliveries from visiting cargo vehicles. That is not ideal, and there's an even greater reason for rethinking a self-generating water solution.

For many years, the station's life support machinery has kept the crew alive by recycling oxygen from water using electrolysis. The hydrogen this produced was considered waste gas and vented overboard. So, too, was carbon dioxide—generated by crew metabolism—vented overboard. NASA knew it forfeited two important consumables but did not have the money to put a more effective system into the baseline for the station. NASA considered a process, originally developed by Nobel Prize-winning French chemist Paul Sabatier in the early 1900s, using a nickel catalyst to interact with hydrogen and carbon dioxide at elevated temperatures and pressures to produce water and methane. The Sabatier process is a well-established water production technology used for many years for advanced military and commercial applications, but the space-based application for the station is unique because of the commercial structure of its implementation.

NASA determined an enhanced Sabatier system could reduce water resupply requirements by thousands of

> Not only was the Sabatier hardware developed and funded by a non-governmental entity, it is operated on a purely commercial basis, with NASA buying the water generated for use on the space station.

NASA astronaut Douglas H. Wheelock, Expedition 25 commander, is photographed with the Sabatier Assembly - Just prior to installation into Oxygen Generator System rack.

Image credit: NASA

pounds of water per year and close the loop in the oxygen and water regeneration cycle. This represented significant and immediate cost savings in the operation of the space station, and provided a way to produce water rather than transport it all from Earth, increasing the goal of self-sufficiency and broadening the path for extended human survival in low-Earth orbit and beyond.

In April 2008, NASA contracted with Hamilton Sundstrand Space, Land & Sea, to provide water-production services aboard the space station that would connect to the existing life-support system. While the company for some time has provided a number of systems for ISS, including spacesuits and those that control electrical power, this agreement expanded its existing work to develop the Sabatier processor. The result was that a 550-pound stainless steel cube the size of a small refrigerator arrived via Space Shuttle Discovery on April 7, 2010, and was operational by October of that year.

The system includes half a dozen major components that ingest, pressurize, condense and transform gases to produce water and methane gas. Besides this production, it is designed for containment of the very reactive hydrogen and carbon dioxide gases. Water is processed through the Water Recovery System. The methane is vented into space, and the water is fed into the station's water system where it undergoes treatment before it is used for drinking, personal hygiene and scientific experiments.

The implementation of Sabatier on ISS is as much about the technological value as about a partnering breakthrough; not only was the Sabatier hardware developed and funded by a non-governmental entity, it is operated on a purely commercial basis, with NASA buying the water generated for use on the space station. The collaboration between NASA and UTC Aerospace Systems has made a significant contribution to the space station's supply chain. Instead of traditional cost-plus contracting roles, the idea was to develop a piece of spaceflight hardware with minimal NASA oversight.

A new business model was in the making. The partners agreed that UTC would fund the development and operation of the Sabatier system, while NASA would launch it on an assembly mission and provide it rack space on the station. Importantly, the agreement removed more than 70 percent of NASA's standard requirements. Verification of the remaining requirements was left as flexible as possible, and specific verification criteria were defined only where absolutely required.

If the system did not work, NASA would owe nothing to UTC; while the system is operational (which it has been for several years now), NASA pays for the amount of water produced. While NASA provided some milestone payments during the development timeframe, these were subject to a 100-percent refund if the hardware did not work upon in-orbit activation. This met NASA's need to keep UTC motivated and met UTC's need for cash flow during the development phase.

ISS is a test bed for exploration, but this illustrates that it can also be a test bed for procurement options. Commercial providers believed that they could deliver a Sabatier system cheaply if the government would just allow them to do so. This experience is serving as a pathfinder for other innovative hardware development on ISS.

Commercialization of low-Earth orbit (LEO)

For too long, in order to utilize the International Space Station and its LEO environment, one had to be an expert, and that presented a significant barrier. Fortunately, an exciting new commercial pathway is revolutionizing and opening access to space—making space just like any other place to do business.

In 2009, NanoRacks started under a very unique Space Act Agreement that enables access to the ISS manifest and in-orbit resources. NanoRacks has self-funded specific research hardware for the station. Utilizing the ultimate "Plug and Play" approach, small payloads in the CubeLab/CubeSat form factor are plugged into platforms or racks, providing interface with space station power and data capabilities. (A CubeSat is a miniaturized satellite that usually has a volume of exactly one liter, or a 10-cm/4-inch cube, and has a mass of no more than 1.33 kilograms/2.9 pounds; the CubeLab is a comparable, compact research environment for inside ISS.)

> An exciting new commercial pathway is revolutionizing and opening access to space, making space just like any other place to do business.

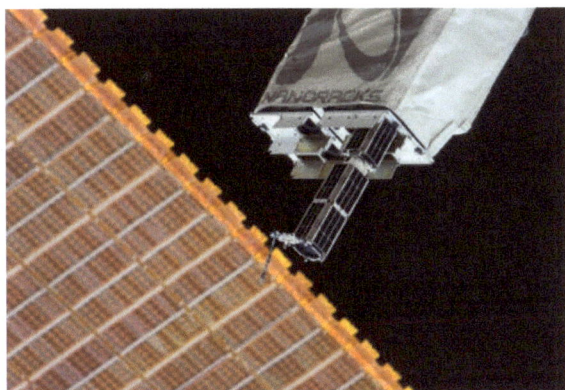

Planet Labs' Dove satellites being deployed from the NanoRacks CubeSat Deployer, Feb. 14, 2014.

Image credit: NASA

In return, the business model enables NanoRacks to market to other organizations as long as it executes the mission of the U.S. National Lab to provide research access to paying customers as commercial users of ISS. This is extraordinary because, for the first time, the market is expressing what can and should be done on the station without direct funding by the government. Over the past three years of operation, NanoRacks has sent more than 200 payloads to station, literally launching a new space market.

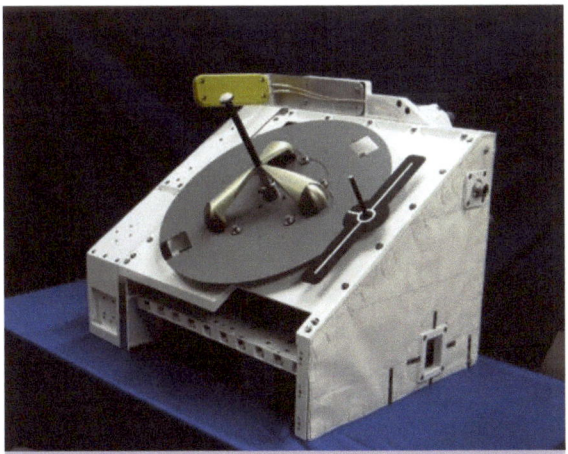

NanoRacks Exposed Platform, scheduled for delivery to the ISS in October 2015 on HTV-5.

Image credit: NanoRacks

Another slice of the new market for NanoRacks is the deployment of small satellites from the station, and no one anticipated its immediate success. NanoRacks saw the opportunity to use the Japanese Experiment Module (JEM) airlock and posed the question to NASA: "If we develop our own more-efficient, less-expensive satellite deployer, can we use it?" The question was not about seeking funding, but permission, to develop and serve a market. NanoRacks identified potential customers on the U.S. west coast working on miniaturization and electronics that wanted to make small satellites but had no way to launch other than through cost-prohibitive Russian or U.S. assets. Subsequently, these companies, as well as labs and organizations throughout the U.S. government, have been very responsive to the NanoRacks CubeSat Deployer (NRCSD). Initially planning to deploy only occasionally, the company now has a queue of customers ready to be deployed from the station.

NanoRacks has applied the power of standardization, the efficiencies of commercialization, and the advances in component miniaturization to in-orbit operations. By focusing on customer satisfaction, they have made it possible for non-experts to use the space environment for experiments.

Private sector participation provides a new model for moving forward in partnership with the government. Through that model, the private sector develops the market, secures the funding, and builds the hardware while the U.S. taxpayer provides the infrastructure and the foundation of the U.S. National Lab in space. The benefits include transparency of costs, low-cost execution, access to space (either for microgravity or the vantage point in low-Earth orbit), speed to market (months as opposed to years), international collaboration and new-idea generation, and broad accessibility over five years.

Innovative public-private partnerships for ISS cargo services: Part 1

In January 2006, NASA announced the Commercial Orbital Transportation Services (COTS) program would be designated to coordinate the delivery of crew and cargo to the International Space Station by private companies. The intent was to spur innovation by the commercial sector to design, build, launch, and fly ISS-destined cargo demonstration flights by September 2013. The companies initially selected for the COTS program were Space Exploration Technologies (SpaceX) and Rocketplane Kistler. However, the companies and spacecraft ultimately completing the

> The Commercial Orbital Transportation Services (COTS) program has been heralded as one of the most extraordinary examples of public/private partnership, and a leap of faith for NASA.

program demonstration requirements were the SpaceX Dragon and Orbital Sciences Corporation's Cygnus vehicles. COTS did not involve binding contracts but did require the successful completion of pre-determined development and financing milestones through the use of Space Act Agreements. A separate program called the Commercial Resupply Services (CRS) was initiated approximately two years after the COTS program began. While the first program developed the transportation vehicles, the second is designed to provide actual cargo and payload deliveries to the station and either cargo return or cargo removal and disposal from the station. The COTS program involved funded Space Act Agreements, with NASA providing milestone-based payments. CRS is a fixed-price services contract, which requires the two suppliers to assume liability for failure to perform their cargo deliveries.

A look inside the decision by NASA to open its doors to private industry is compelling.

Station resupply and disposal was the first capability area—requiring precision orbit insertion, rendezvous, and docking with another spacecraft—with commercial crew transportation to follow.

Today, with space station cargo resupply efforts underway, it is clear the unprecedented efficiency of the COTS investment resulted in two new automated cargo spacecraft. It has been heralded as one of the most extraordinary examples of public/private partnership, and a leap of faith for NASA.

SpaceX's achievements include the first privately funded, liquid-fueled rocket (Falcon 1) to reach orbit on Sept. 28, 2008; the first privately funded launch, orbit and recovery of a spacecraft, Dragon, on Dec. 9, 2010; and the first private spacecraft (Dragon) to launch to the station, on May 25, 2012. As of February 2015, SpaceX has flown six cargo missions to the ISS.

Founder Elon Musk focused the vision early, utilizing a vertical integration business model—an arrangement in which the supply chain of a company is owned by that company. In order to control quality and costs, SpaceX designs, tests and fabricates the majority of its components in-house, including the rocket engines used on the Falcon launch vehicles and the Dragon spacecraft. This type of production is unusual in the aerospace industry but has allowed SpaceX to significantly reduce conventional rocket development and flight integration time. In addition to the capability of delivering pressurized cargo, the Dragon vehicle also has a "trunk" allowing the transport of unpressurized payloads intended for the exterior locations on the station (and for disposing of them once their operational life or research mission is complete). Perhaps most importantly, it can return cargo and experiment samples to Earth in the pressurized volume. This

SpaceX's Falcon 9 lifts off with its Dragon resupply vehicle aboard, headed for the ISS.
Image credit: NASA

SpaceX's Dragon capsule as it approaches the ISS, Oct. 25, 2014.
Image credit: NASA

> The new resupply services contracts represent a major change in the way NASA procures space transportation services.

return of recoverable capability had been all but lost to NASA since the retirement of the shuttle; the only other means of returning very small amounts of pressurized cargo has been with the crews as they return to Earth via Russian Soyuz vehicles.

During a keynote speech in 2013 at the International Symposium for Personal and Commercial Spaceflight (ISPCS) Conference, SpaceX Chief Operating Officer Gwynne Shotwell described the notable change in public perception about space. She said the surge in commercial opportunities through partnerships with NASA and the rebirth of entrepreneurial organizations has spurred renewed awareness and excitement, including a growth in technical and engineering jobs.

Innovative public-private partnerships for ISS cargo services: Part 2

An important part of the COTS story is that a human-inhabited space station presents a useable, continuous market for commercial ventures. Initially, there was a crucial need for cargo delivery to the station. The COTS and CRS programs enabled a new mindset about resupply missions, with cost and assumption of risk no longer weighted on the government and the opportunity for private contractors to propose innovative solutions. The new resupply services contracts represent a major change in the way NASA procures space transportation services.

Orbital Sciences Corporation (Orbital) has been involved in the commercial use of space for 33 years. Orbital sells satellites commercially to operators around the world. That business model evolved to incorporate rocket development and integration at the launch site. Having been a partner with NASA for many years on multiple projects, the COTS and CRS projects allowed Orbital to be actively involved in the ISS program, NASA's premier human spaceflight endeavor, leveraging its spacecraft and rocket experience to date. Additionally, it inspired Orbital to invest in more research and development to support the new endeavor.

Orbital's Antares rocket with the Cygnus cargo vehicle aboard, launches to the International Space Station.
Image credit: NASA

Orbital's Cygnus vehicle about to be berthed to the International Space Station, July 16, 2014.

Image credit: NASA

As one might imagine, the level of financial risk for a commercial company in the rocket and satellite business is greater than other means of transportation. There is always potential for failure. A case in point is the total loss of the Orbital 3 flight immediately after liftoff in October 2014. The loss of the cargo destined for the station required all parties to work together to re-plan and recover from the unfortunate event. Publicly traded Orbital had built cost control into the structure of the company. The books are open for the Security Exchange Commission and stockholders to see, so there has to be transparency in company finances. Insurance companies are incorporated into the mix to help manage risk. Another challenge is that customer requirements tend to change and there is a desire to be flexible, but changes cost money. Past systems such as the space shuttle used cost-plus contracting. If NASA sought to change a requirement, the contractor complied and issued an additional bill. Under the fixed-price contracting of today, the change request is reviewed by all parties for potential scope increase and evaluated for the additional cost. Alternative in-scope versus out-of-scope ideas, called "what if" scenarios, can be brainstormed without incurring a lot of cost because of the collaborative nature of all involved. Competition also makes the parties more cognizant of the costs.

Commercialization of various aspects of NASA's mission is having a very positive ripple effect. New companies participating in human spaceflight and developing unique partnerships has given regional economies a boost. Multiple international suppliers have close ties to their respective space agencies, which enables more collaboration and continuity. And large and small U.S. aerospace companies now have the ability to contribute and join these ventures.

With SpaceX and Orbital leading the way in carrying out the resupply service at a lower cost to the taxpayer, NASA can give its attention to other space station and exploration objectives.

Precision pointing platform for Earth observations from the ISS

In May 2014, the German Aerospace Center (Deutsches Zentrum für Luft- und Raumfahrt; DLR) and the U.S. corporation Teledyne Brown Engineering, Inc. (TBE) announced an agreement to install and operate the imaging spectrometer DLR Earth Sensing Imaging Spectrometer (DESIS) aboard the International Space Station. As one of four DLR-built camera systems for remote sensing, it will be fitted to the Multi-User System for Earth Sensing (MUSES) precision-pointing platform TBE is developing for the station. DESIS will be able to detect changes in the land surface, oceans and atmosphere—contributing to a better understanding in fields as diverse as ecology, agriculture, and urban land use. The project is intended to open up new possibilities in Earth observation and is one of the first to be developed for ISS in cooperation with a commercial enterprise. This follows by just two years TBE's award of a Cooperative Agreement by NASA to foster the commercial utilization of the space station by developing MUSES, which in turn launched the company's new commercial space-based digital imaging business.

MUSES is a platform designed to host Earth-looking instruments, such as high-resolution digital cameras and multi- or hyperspectral instruments, and provide precision pointing, inertia-stabilization, and other accommodations.

The new platform is about a meter square, with gyros, star trackers (small aperture, space-qualified optical

> We must establish enough momentum to sustain commercial enterprise in low-Earth orbit when the current station is no longer available.

products that ensure a spacecraft's accurate attitude in space), and step motors (used for robotics and handy devices for repeatable positioning) designed to keep that platform fixed in inertial space, so as to negate the wobbles and vibrations of the station, which is critical for image quality. MUSES can host up to four instruments simultaneously to look down at the Earth. The model for commercialization of the station is evolving such that the company contributed its own resources while NASA shared in the development costs as well as providing the launch and in-orbit infrastructure. TBE has secured a National Oceanic and Atmospheric Administration (NOAA) commercial imaging/remote-sensing license so that it can operate instruments on MUSES and then sell the data commercially. The company likens the station to a piece of real estate several hundred miles up in the air that is being developed over time for scientific, economic, educational, commercial, and quality of life purposes.

The first instrument placed aboard will be the DLR's DESIS hyperspectral instrument, which has the capacity to distinguish slight variations in the reflectance of sunlight from the Earth surface (in the visible through near-infrared spectrum) when pointed over a geographic area. An image spectrometer is able to distinguish very subtle changes in the reflectance spectrum for distinguishing plant species or whether the forest is undergoing some sort of stress due to drought or pests. Fine variations in surface reflectance can give immense amounts of information just not possible with picture, such as you would take with an off-the-shelf digital camera.

Remote sensing of the Earth has come a long way, driven by concerns about agriculture, oil or gas resources, biodiversity, mineralogy, change detection and monitoring of world heritage sites, coastal zones, water ecosystems, and transportation. The DLR/TBE commercial enterprise introduces a very specific measurement tool that will improve with application, for many markets that already exist. It is as much about the uses of the data as the data itself. For example, if a farmer wants to know when, where and how to water his corn, he doesn't really need to know all the technology behind a spectral library.

TBE was motivated to think more commercially about how it does business. TBE responded to a broad agency announcement (BAA) with an idea for putting a platform on the station and starting a commercial business leveraging its existing infrastructure. NASA and TBE recognized that the quickest way to extract value from the ISS would be to bring down data.

TBE knows it is at one end of the pipeline—its users will benefit from the use of the data—and by providing data that is otherwise not available for application in industries such as fisheries and agriculture, TBE will be testing a commercial model for other instruments in the future.

Visioning beyond the space station will be critical for NASA and companies like TBE. To sustain what is now a nascent commercial marketplace in low-Earth orbit, participants must understand, accept, and plan for the day when the station will no longer be there. We

The MUSES facility on the ISS will provide a state-of-the-art pointing platform for Earth observations.
Image credit: TBE

The MUSES platform undergoes final testing at the TBE labs, Huntsville, Ala.
Image credit: TBE

must not only leverage and develop the commercial market utilizing current assets but smoothly portage that whole commercial economy from the current station to whatever other infrastructures we can foresee, will build, or have available in LEO or beyond. In other words, we must establish enough momentum to move the commercial enterprise beyond when the current station is no longer an asset at our disposal. Though this might seem difficult, it actually energizes the foundations on which our country was established: the challenge to persist, to contribute, to open new possibilities, and to make this world a better place.

The Groundbreaker: Earth observation

In 2014, NanoRacks signed an agreement with a publicly traded Canadian company called UrtheCast to place Earth Observation (EO) sensors on the American side of the space station, having already placed two instruments on the Russian side. UrtheCast's entire business model is tied now to the ISS. What's more, through this partnership, UrtheCast is making one of the single largest private investments to date in utilization of the station.

The genesis of a new company began four years ago when MacDonald, Dettwiler and Associates (MDA), Canada's largest aerospace company, explored options for placing Earth observing sensors on the Russian segment. Ultimately, MDA determined the project was not compatible with its corporate mission, so did not pursue it further. However, a couple of executives took the opportunity to spin the company out and pursue the prospect directly, calling the new firm UrtheCast. They developed an experiment to build and operate two cameras on the ISS and market the imagery worldwide.

Both cameras launched in November 2013 and were installed in January 2014 via two spacewalks by Russian cosmonauts. With two sensors in place, UrtheCast went on to raise $68 million privately and ultimately executed an Initial Public Offering (IPO) on the Toronto Stock Exchange.

About the same time the two cameras were installed, UrtheCast started actively seeking a U.S. partner, ultimately partnering with NanoRacks, seeking its access to station. It was a natural fit, which resulted in a June 2014 announcement that a suite of next-generation sensors would be developed and ready to launch in late 2016 or early 2017.

The UrtheCast/NanoRacks partnership plans to put its own small module, called Lightweight UrtheCast NanoRacks Alcove (LUNA), on the U.S. side of station. The proposed project would contain three elements:

1. A very high-resolution, dual-mode and multi-spectral camera, capable of switching between still images and high-resolution video. Developers hope that after processing, the sensor can achieve resolution as low as 40 centimeters, which would make it one of the highest-resolution, Earth observation instruments in orbit.

2. A Synthetic Aperture Radar (SAR) sensor, which will send an energy pulse down to Earth, providing the ability to see at night and through clouds. The data set is not intuitive to look at and very difficult to handle, but fusing the data sets ultimately creates a cloudless image.

3. The two sensors will be paired with a dedicated X-band downlink and distributed through Internet Cloud technology, enabling a capability akin to Google Earth and Netflix.

Urthecast camera being installed on the Russian Segment of the International Space Station.
Image credit: Roscosmos

> UrtheCast is a company whose entire business model is tied to the ISS and is making a large private investment to use the station.

Captured by UrtheCast's Theia camera on July 3, 2014, this image spans across Rome, Italy, and its surroundings. Image credit: UrtheCast

The station is a very capable imaging platform. It is in low-Earth orbit, which means very high-resolution instruments can provide great images, its orbit covers approximately 95 percent of the Earth's inhabited areas, and it has the ability to revisit locations regularly.

The partnership between UrtheCast, NanoRacks, and NASA allows instruments to be put in orbit at a lower cost by benefitting from existing infrastructure and investment. These companies believe that dreams for space exploration will remain politically and economically difficult to achieve unless an energized customer base emerges. They believe that data from state-of-the-art EO sensors on the station offer an obvious solution to help stimulate that demand.

A flock of CubeSats photographs our changing planet

Around 10:50 a.m. on July 23, 2014, the California Highway Patrol received a report of a fast-growing wildfire in Riverside County, west of Palm Springs. The blaze spread quickly, forcing some residents to evacuate. Within 15 minutes, a toaster-sized satellite called a Dove captured an image depicting the fire's size, the path it burned through, the wind's direction, and the fire's exact location. The fire was fully contained within a day. The satellite took the picture without anyone requesting it, simply performing its normal task of photographing a 10-kilometer by 15-kilometer strip of Earth (about 6 miles by 9 miles) every second, a job enabled by the International Space Station.

Larger government and commercial satellites provide nations and corporations with detailed images of the planet, some of which people can browse to look for their own neighborhoods. These satellites are expensive and limited in number, though, and either have low spatial resolution or take images according to customer requests, focusing only on certain areas. Crucially, these satellites might not take repeat images of the same region for several months. This makes it difficult to study rapid changes, such as international conflicts, environmental degradation or fast-growing forest fires.

A small-satellite startup called Planet Labs is using the space station as a launch pad for a fleet of miniature

> Planet Labs aims to make Dove images searchable and easily available for everyone from governments studying changing shorelines to firefighters battling a forest fire.

Two Planet Labs "Dove" CubeSats immediately after being deployed from the ISS.
Image credit: NASA

orbital cameras. These "flocks" of 11-pound (5 kg) CubeSats, or Doves, take continuous photos of the entire Earth and transmit them to ground stations every 90 minutes. They provide an unprecedented view of Earth, from growth in cities to summer wildfires like the one in California.

Different types of CubeSats frequently launch from the space station and provide a novel way to image Earth, log ship traffic, and send messages, among many other tasks. An entire flock allows Planet Labs 8 to collect planet-wide data, which would be impossible with a larger single satellite. The company started in Silicon Valley in 2009 with a goal of capturing a daily global mosaic of Earth. This requires a large fleet of CubeSats, and the station is the key to getting them into space.

From its perch 230 miles (370 km) above Earth, each Dove captures images once per second and each has a resolution of 10 to 16 feet (three to five meters). That means each pixel in the camera's viewing area corresponds to a 3-meter-wide object. That resolution is enough to distinguish individual trees in a rainforest but not enough to identify a person walking on the street. Planet Labs has launched 71 Doves since January 2014 using CubeSat deployers developed by NanoRacks, LLC. In part because the CubeSats are so small and relatively inexpensive, they have a lifespan of only a few months of operations, but their orbital altitudes are low enough that they eventually fall back to Earth and, by design, burn up in the atmosphere. Doves are continually replaced by newer models, which Planet Labs is constantly testing and updating from its San Francisco headquarters. To date, Planet Labs has built more than 100 Doves. The company is funded through private venture capital and private investors.

Eventually, Planet Labs aims to make Dove images searchable and easily available for anyone who wants to look at them—from spectators eager to see their houses, to governments studying changing shorelines and firefighters battling a blaze in a small slice of forest.

Stretch your horizons, Stay Curious™

Kentucky Science and Technology Corporation (KSTC), and two subsidiaries, Kentucky Space (a nonprofit company) and Space Tango, Inc. (a for-profit enterprise) are building a reputation for designing and executing diverse initiatives in science, technology, entrepreneurship and disruptive innovation. The focus is on small, high-value satellites and applications, and novel space platforms and experiments for the space station.

Kentucky Space began as a vehicle for involving university students in space projects. It is an ambitious consortium of universities and public and private organizations aimed at designing and leading innovative space missions within realistic budgets and objectives. The programs cover a spectrum of flight opportunities including near-space balloons; sub-orbital, orbital and

Irrigated fields in Pinal County, Arizona, as viewed by a Dove CubeSat.
Image credit: Planet Labs

> Space has always been at the forefront of innovation and really pushes any company to challenge the status quo.

ISS missions; as well as partnerships with organizations and space agencies worldwide. Twyman Clements was a student at Kentucky Space only a few years ago; now he is the young CEO of Space Tango.

Based in Lexington, Space Tango seeks to utilize space to discover, design and commercialize solutions for applications on Earth. This is accomplished through platforms—hardware and software offerings to those in the research communities looking to utilize the unique environment of microgravity. Focused primarily on the entrepreneurial space marketplace, Space Tango's capabilities and experience involve CubeSat class and other micro-satellites and subsystems, satellite ground operations, space platform design and testing, and development of novel technology and experiments for the station. The company is committed to a highly collaborative business strategy and works closely with a number of other companies, universities and organizations.

A general research platform/multi-lab facility, called TangoLab by Clements, will be installed in 2015 on the ISS. It will host a variety of payloads relevant to the medical field, the material sciences, and environmental sensing. Researchers will interact with their experiments in the unique space environment right from a computer, with data transmitted over the internet.

The CubeLab arose from a need for repeatable/affordable research on the station. The uniquely designed research module was developed by Kentucky Space, based on the proven CubeSat form-factor. Providing a standardized platform and open architecture for the experimental modules shortens the development cycle and lowers costs for research and development. The goal is to lead to quicker time to market for new drug products, treatments or procedures.

The TangoLab research platform is a work of elegant art. It was not simple to develop, requiring the collaborative experience and creativity of everyone involved, with the aid of 3-D printing. The premise, though, is simple: use of standardized hardware dramatically lowers the cost of space station utilization.

In May 2011, Kentucky Space launched an exciting new initiative: the study and exploration of medical solutions in the microgravity environment of space. The Exomedicine Institute, an interdisciplinary team of top scientists, engineers and entrepreneurs, has been assembled to forge ahead into this new and potentially promising field of research.

All organisms on Earth, during evolution over billions of years, have adapted their form and function to the force of Earth's gravity. These characteristics are encoded in their genes: up-down asymmetry, structural strength, size of force-producing elements and sensory systems.

Since gravity influences all biological systems at a molecular level, what happens to biological systems when the influence of gravity is modified or removed? We can't remove gravity on Earth, but conducting biomedical research in space makes certain Earth-bound limitations disappear.

Without gravity, cells, molecules, protein crystals, and microbes behave in very different ways. Microgravity thus presents opportunities to explore new and potentially game-changing discoveries in areas such as human tissue regeneration, drug development, treatments for diseases such as cancer and other life-threatening and chronic conditions, as well as energy and novel materials.

The work of Kentucky Space, Space Tango, and the Exomedicine Institute reflect an interdisciplinary approach to research and development (R&D), as well as the rise of a "newspace" era with a multitude

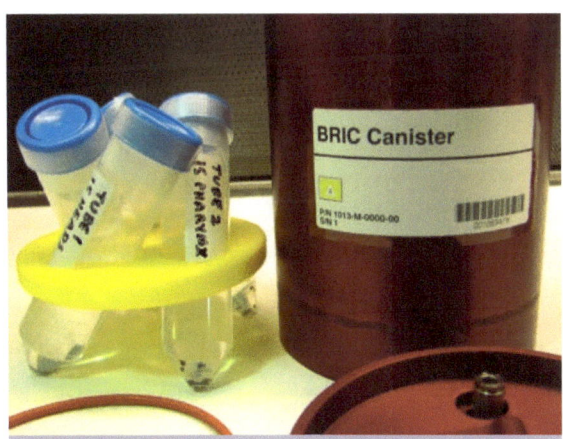

Flatworms being prepared for loading into the Biological Research In Canisters hardware.
Image credit: Space Tango

of players. Lowering barriers enables people to do research in, and manufacture for, LEO. Based on a full-service philosophy, these three companies recognize that the average person knows little about handling a launch with NASA, so the companies offer engineering and technical expertise to assist with experiment design, payload design and integration, facilities, and a full suite of test equipment and capabilities for a successful launch.

Going into space forces people to think outside the box; space has always been at the forefront of innovation and really pushes any company to challenge the status quo.

In 2000, Kimel created another entity as well: IdeaFestival, an international event centered on innovation, discovery and creative thinking across different disciplines. IdeaFestival is an annual four-day gathering of an eclectic network of global thinkers and one-of-a-kind innovators bound together by an intense curiosity about what is impacting and shaping the future. Those who attend leave the event with interesting new ideas, a better sense of connectedness, an expanded network of relationships and lasting inspiration to help create change in the world.

These types of opportunities help inform others of new ideas, fueling the pipeline of newspace activity with a vision to increase private sector entrepreneurship through student pioneers.

As the IdeaFestival mantra goes: Stretch Your Horizons, Stay Curious—for a new idea can be generated when you least expect it.

Mission critical: Flatworm experiment races the clock after splashdown

Everyone is familiar with the blastoff countdown associated with a rocket headed to space, but what about the countdown that takes place once a capsule returns to Earth?

The highly sensitive biomedical experiments that travel aboard the International Space Station have the potential to impact lives. Success of these experiments often depends on a swift shipment from the splashdown site to the lab, where scientists can analyze crucial data. The moments between splashdown and analysis are critical.

This was the case when a capsule from the station carrying 48 planarian flatworms (known for their ability to regenerate their own body parts) splashed down in the Pacific Ocean off the coast of California in 2014.

The flatworms had been delivered to the space station one month earlier as part of a collaboration between independent, nonprofit space contractor Kentucky Space and its for-profit arm Space Tango, the Center for the Advancement of Science in Space (CASIS), NASA and the Tufts Center for Regenerative and Development Biology. The experiment could lead to powerful benefits for humanity, including further understanding of the regenerative powers of the flatworms that could someday impact global healthcare, including treatments for cancers, treatment of spinal cord injuries, and the ability to correct developmental problems in embryos.

Two factors were critical to the success of the research: time and temperature. The flatworms needed to be transported quickly from Long Beach, California, to Tufts University in Boston for analysis, and the shipment needed to maintain a temperature of 12

> Innovative logistics by FedEx enabled the possibility to connect the space station and a strong terrestrial team to advance research.

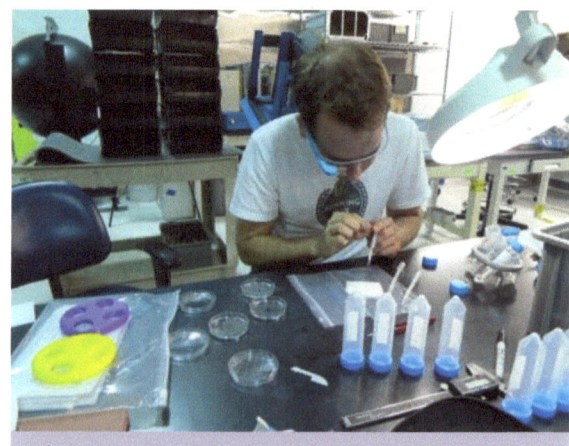

Flatworm specimens are prepared prior to their launch to the ISS.
Image credit: Space Tango

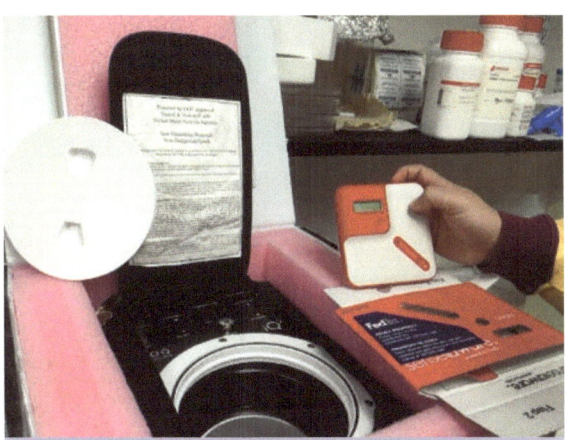

This FedEx environmentally controlled shipping container, with SenseAware environmental monitoring, transported the Space Tango flatworms experiment to the lab after returning to Earth from the ISS.

Image credit: FedEx

Economic development of space in JAXA

Japan Aerospace Exploration Agency (JAXA) has been striving for economic development of space in the Japanese Experiment Module (JEM, or "Kibo," which means "hope" in Japanese) and H-II Transfer Vehicle (HTV, or "Kounotori," which means "white stork" in Japanese). Economic development of space includes advancing technology of Japanese companies, expanding spin-offs and opening the door to commercial sectors in Kibo utilization.

Navigation and Communication Services for the International Space Station

Throughout the history of the ISS program, many Japanese industries have been challenging traditional approaches to human spaceflight and acquiring advanced human space technologies.

Some Japanese firms, with acquired globally competitive technologies and know-how, have delivered human space technology systems to the worldwide commercial market.

degrees Celsius throughout its journey. Mishandling of either environmental factor could kill the flatworms and cause the experiment to fail.

FedEx was up for the challenge. Space Tango, on Kentucky Space's behalf, worked with the FedEx Space Desk team, logistical experts for the space industry and the power behind FedEx® Space Solutions, to design the optimum journey for the flatworm experiment.

When the capsule hit the water, the countdown began.

The vessel was rapidly retrieved and the flatworm shipment quickly sent on its way to Boston via FedEx Express, inside a custom-designed, temperature-controlled package. The shipment was monitored throughout its journey by SenseAware, powered by FedEx, which allowed stakeholders at Kentucky Space, Space Tango and Tufts University to not only track its location, but also remain certain that the temperature inside the package did not waver from 12 degrees Celsius.

The shipment arrived safely and on time in Boston, giving Tufts' scientists near-immediate access to the flatworms. Mission: Accomplished.

Research continues on the effects of a microgravity environment on the regenerative ability of flatworms. Innovative logistics by FedEx enabled the possibility to connect the space station and a strong terrestrial team—all of which played a key role in advancing the research in this important field.

> There are many undiscovered possibilities for increasing space station utilization. JAXA already has discovered and successfully accomplished a number of commercial projects aboard Kibo.

An example is the Proximity Link system applied for commercial spacecraft. Mitsubishi Electric (MELCO) designed and produced the Proximity Link System (PLS), which communicates with Proximity Communication System (PROX) installed in the Kibo module, providing navigational information and communication links to space vehicles to safely approach and depart from the station. MELCO originally designed and produced this PROX/PLS to enable HTV, or Kountori, to rendezvous and berth with the station. This brand new technology has since been applied to Orbital's Cygnus spacecraft to enable its safe arrival at the station.

Commercial Utilization Initiatives

JAXA has endeavored to utilize the Kibo module commercially for research by following three core initiatives based on our human space technologies and experiences of its utilization: promoting use of the station to new potential commercial markets, creating new opportunities for commercial customers to use the station, and enhancing the capabilities of Kibo.

Promotion to Potential Commercial markets

JAXA searched promising research areas with ground applications and recognized the potential benefit for drug design using protein crystal growth experiments, Earth observation and generation of new materials.

For Earth observation, the Japan Space Forum delivers the high-quality movies of the Earth via internet with its Kibo Hi-Vision EarthView Educational Program service.

The possibility of generating new materials in space has attracted commercial companies. One example is the colloidal crystals as novel optical materials. Large, high-quality colloidal crystals made in space are expected to have applications for photonic materials. This experiment was executed as a collaborative work between Hamamatsu Photonics K.K., Fuji Chemical Co., Ltd. and universities.

The Production of High Performance Nanomaterials in Microgravity (Nanoskeleton) investigation aimed to clarify the effect of gravity on oil flotation, sedimentation, and convection on crystals generated in microgravity. Nanoskeleton data will be added into a computational chemistry simulation for Nanoskeleton synthesis, and the simulation will be used for the prediction of the proper parameters for synthesis on the ground. Shiseido Co., Ltd. and universities took part in this project.

Strategic promotion by JAXA has increased the number of commercial customers, including non-traditional users. One example is drug companies benefitting from the protein crystal growth experiments in Kibo. Another example is space experimentation on the effects of probiotics on immune function, a collaborative work between a beverage company, Yakult Honsha Co., Ltd. and JAXA. Astronauts may be at physical risk because a long stay in space is known to alter human immune function. A probiotic, *Lactobacillus casei* strain Shirota, has been demonstrated to offer health benefits by improving intestinal microbiota and maintaining immune function on the ground. An investigation in Kibo examined the health benefits of probiotics consumption on human immune systems aboard the space station. It may lead to the development of functional space food and techniques for long-term Lactobacillus preservation on Earth.

New Opportunities for Commercial Customers

In 2013, JAXA started a new commercial utilization scheme, based on knowledge from previous activity,

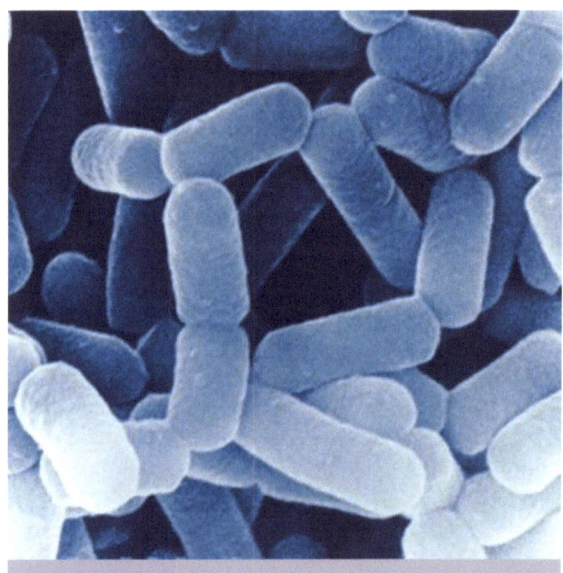

Lactobacillus casei strain Shirota.
Image credit: Yakult Honsha Co., Ltd.

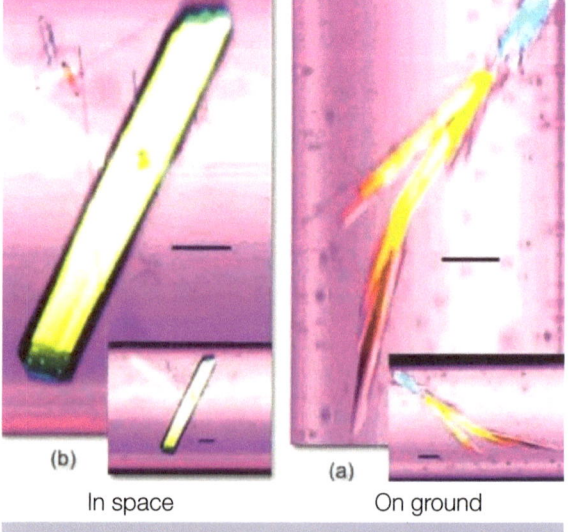

Protein crystal growth for drug development.
Image credit: JAXA

to provide more user-friendly utilization services and a system to facilitate entry and shorten the time required for implementation of space activities. JAXA's efforts to encourage new entries from non-traditional users into the space field continue.

One of the major services already available to commercial users is the JEM Small Satellite Orbital Deployer (J-SSOD), which deploys small satellites for Earth observation, remote sensing and capability building. J-SSOD is an excellent example of producing new business using an existing facility, combining the JEM, airlock and robotic arm. This conceptual breakthrough has spread. In fact, once the deployment of small satellites using the airlock had proven to be a valuable capability, a US company—with NASA as a customer—developed their own deployer to take advantage of this opportunity.

A Brazilian microsatellite, AESP-14, was successfully deployed in February 2015 from Kibo at Tsukuba Space Center. AESP-14 was developed by Technological Institute of Aeronautics (ITA) with support of Brazilian Space Agency (AEB) and Brazilian National Institute for Space Research (INPE), with Japan Manned Space Systems Corporation (JAMSS) ensuring deployment from Kibo. This was the first opportunity for commercial utilization of J-SSOD.

Various new space experiment technologies are being developed such as crystallization technique that acquires large single crystals for neutron diffraction and widening temperature range to meet commercial users' demands. Protein crystal growth project has attracted research-and-development oriented drug companies through targeted direct promotion. Interprotein Corporation, Chugai Pharmaceutical Co., Ltd. and ARKRAY, Inc. participated in JAXA's protein crystal growth experiments. Interprotein set out to achieve high-quality, co-crystal structure of proteins and low-molecular compounds for effective design of drugs. Chugai aimed at precise 3-D structures of proteins by the high-quality crystals grown in microgravity, to help understand the structure/function relationship of drug candidates and create revolutionary new drugs. ARKRAY and Tokyo University of Agriculture and Technology took part in JAXA's experiment to analyze the protein structure that is indispensable for the development of biosensing technology and aim at its application for an innovative biosensing system that will be helpful for the treatment and diagnosis of diabetes.

Enhancing the Capabilities of Kibo

JAXA continues upgrades of existing Kibo facilities as well as new facility installations, including the

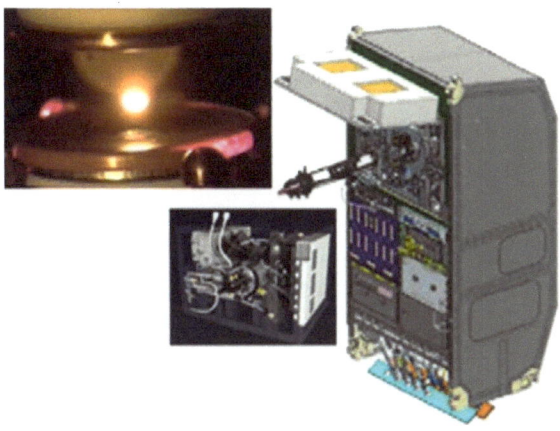

Electrostatic Levitation Furnace (ELF) for new material generation and high-temperature thermal property data acquisition.
Image credit: JAXA

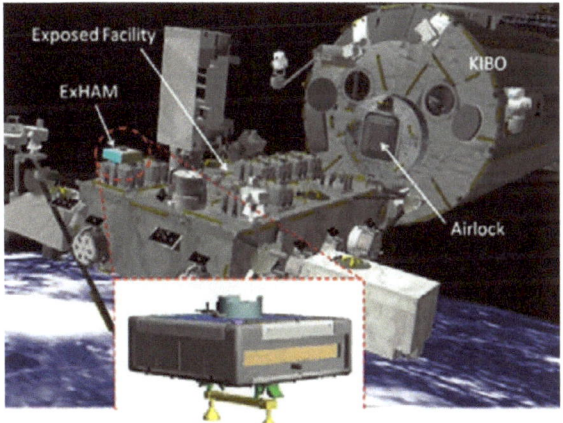

Experiment Handrail Attachment Mechanism (ExHAM) attached to the JEM Exposed Facility.
Image credit: JAXA

Rodent facility (Mouse Habitat Unit) for rodent research, fluorescence microscope and Electrostatic Levitation Furnace (ELF) for new material generation research and high-temperature thermal property data acquisition.

The space station also serves as a test bed for future exploration, satellite technology validation and transfer vehicle technology development. JAXA provides simple external experiment opportunities for technical demonstrations. For example, the Exposed Experiment Handrail Attachment Mechanism (ExHAM) enables exposure of materials to the space environment outside the station, and cosmic dust extraction. Junkosha Inc. will utilize ExHAM facility to expose the PEEK electric wire material, which has high-temperature and high-radiation resistance, to the space environment. Acquired data will be evaluated for the company's wire commercialization in space craft use.

There are likely many undiscovered possibilities for increasing space station utilization. JAXA already has discovered and successfully accomplished a number of commercial projects aboard Kibo and will keep finding innovative ways to facilitate its commercial utilization. This will benefit all humanity as well as provide benefits to Japan.

Commercial Research

The unique environment of microgravity provides opportunities for many types of commercially-viable research. Using model organisms (such as rodents or flatworms) to help understand terrestrial concerns such as bone loss or muscle wasting, performing materials research on colloids to develop products that are more uniform and have a longer shelf life, growing larger protein crystals on the space station to help develop monoclonal antibodies, and using the station as a launchpad for a flock of Earth-observing satellites, are just a few examples of the diverse research interests of the corporate world and how they intersect with the International Space Station. These summaries of commercial research activities in progress show the impact and interest in using the space station for research and development.

Colloids in space: Where consumer products and science intersect

The Proctor & Gamble story represents two types of people: the consumer and the materials scientist.

As a rule, consumers want to use products that enhance their lives. Those products must fulfill the promises they promote, whether to have a long shelf life, be easy to use, or perform as advertised. In the case of products such as shampoos or liquid soaps, the purchaser does not want to see obvious physical separation of the material within the bottle (a sediment or settling), which could indicate something amiss.

Materials or physical scientists, for their part, want to produce products or discover formulas that hold new promise and function. They want to know how certain active ingredients or stabilizers behave when added to product formulas. They constantly hunt for the optimal mix that will increase product shelf life or performance. Sometimes it takes working under very special circumstances, which is why materials scientists at Procter & Gamble Co. (P&G) decided to take their research to a higher level—aboard the orbiting laboratory of the International Space Station. Working with NASA, P&G has funded an investigation of how gas and liquid phases separate and come together in microgravity in the study of colloids.

On Earth, gravity complicates this research by causing heavy components to sink and lighter ones to float. This movement occurs very quickly, making it difficult to understand what is happening and why. Space, however, negates these gravitational forces, revealing the natural movement of the colloids. The in-orbit samples' aging process works more slowly and evenly, making it easier to study.

> Using microgravity to study the exceptionally small particles that make up liquid products may ultimately lead to improving health, beauty and household care products that we use every day.

A series of colloid investigations aboard the space station were conducted from October 2013 to March 2014 using a science platform known as the Advanced Colloids Experiment (ACE), it is a collaboration between P&G and Case Western Reserve University.

Though the investigation is designed to help researchers understand how to optimize stabilizers for extending product shelf life, the results also are intended to cut development, production and transportation costs. Better stabilizers result in better quality, reduced costs and greener, more concentrated products that use less plastic packaging, resist collapse, and remain consistent throughout their life. In such an improved process, the first ounce coming out of the bottle will be the same as the last.

Driving P&G's interest in research on the station is the fact that about two thirds of its biggest brands are soft-matter systems—things like fabric softener, deodorant and detergent—that could benefit from this study. Armed with a better understanding of the nature of the

 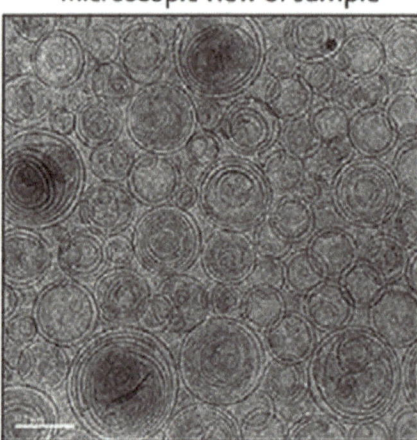 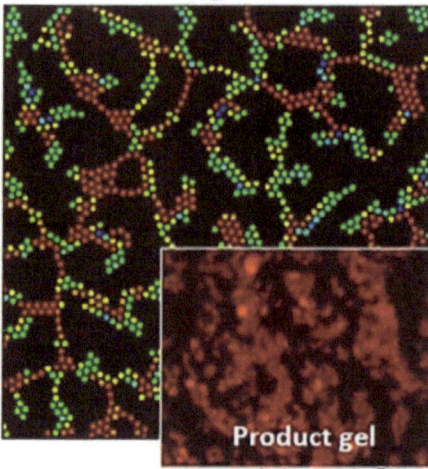

Similar to what is under investigation in the Advanced Colloids Experiment, the above microscopic view of product sample gel in microgravity is dominated with fragile strands composed of many particles in a cross-section. The Model of gel structure above reveals characteristics hidden by gravity.

Image credit: NASA

fluids separation, researchers can work on creating better stabilizers and product formulations.

ACE-M-1 studied behavior of microscopic particles in liquids, gels and creams. ACE-M-2 is currently operating on the space station and continues work in phase separation and how to influence phase separation. The investigation examines the behavior of model colloid-rich liquids and model colloid-poor gases near the critical point, or the point at which no distinct boundary exists between the two phases. ACE-M-2 records micro-scale events on short time scales, while previous experiments observed large scale behavior over many weeks. Liquids and gases of the same material usually have different densities and so would behave differently under the influence of gravity, making the microgravity environment of the International Space Station ideal for these experiments.

The ACE-M-3 experiment involves the design and assembly of complex three-dimensional structures from small particles suspended within a fluid medium. These so-called self-assembled colloidal structures are vital to the design of advanced optical materials. Researching them in the microgravity environment will provide insight into the relation between particle shape, crystal symmetry and structure: a fundamental issue in condensed-matter science.

One reason P&G is involved in this research is its interest in understanding phase separation kinetics of colloids, masked by gravity on Earth. The small blobs that form and grow in microgravity, instead of

European Space Agency astronaut Paolo Nespoli operates the Light Microscopy Module microscope aboard the International Space Station on a previous mission.

Image credit: NASA

the simple top and bottom phase seen on Earth, allow recording of the kinetics of this process. The scientific insights that result from having this data available can lead to more efficient and improved product formulations that are less expensive to produce and/or provide longer shelf life. For a product such as Downy, with sales of about $4 billion a year, even a small one percent savings in production costs or a slightly longer shelf life provides a significant return on the investment. P&G will spend $10 million this year on research to address product shelf life problems.

The results of this work can impact far more than fabric softener.

The potential for the ACE research application in other areas is something that continues to grow. For instance, advanced colloidal formulas also could conceivably lead to improvements in items such as liquid pharmaceuticals, which can be ineffective or even dangerous if not properly mixed when consumed. With better formulations, consumers could look forward to the certainty of a perfect product every time.

Gaining a better understanding of the physical processes of particles obtained through ACE samples, for example, may greatly impact the quality, production and longevity of commercial products. Using microgravity to study the exceptionally small particles, known as colloids, which make up these types of liquid products, researchers can gain more insight into their characteristics. This may ultimately aid research efforts in improving products people use every day.

Space mice teach us about muscle and bone loss

Bone breakage, buildup and eventual loss have a significant impact on our bodies. Bone loss occurs at an accelerated rate in space because of the lack of normal weight-bearing activities in the microgravity environment. Using nutrition and specific exercises, the crew aboard the International Space Station can partially mitigate these concerns. This accelerated aspect of bone loss in spaceflight provides an opportunity for researchers to identify the mechanisms that control bones at a cellular level. While most people will never experience life in space, the benefits of studying bone and muscle loss aboard the station has the potential to touch lives here on the ground.

Bone loss from osteoporosis is a major concern for the elderly. However, inactivity from injury, illness, or malnutrition from anorexia or dietary challenges can lead to bone breakdown in otherwise healthy people. For some time, researchers have tried to understand this phenomenon and have looked at rodents flown aboard

A view inside of the NASA Rodent Research facility, self-contained habitat that provides its occupants living space, food, water, ventilation, and lighting. Cabin air is exchanged with the facility, creating a slight negative pressure inside the cage to pull animal waste into a collection filter.

Image credit: NASA

> Now that the station is complete, rodent research can continue for longer test runs than shuttle missions permitted, thereby increasing data collection and the potential for discovery.

space shuttle missions to the space station in a series of experiments using the Commercial Biomedical Testing Module (CBTM).

The CBTM studies came about as a result of two former Bioserve graduate students going to work for Amgen, a California-based biopharmaceutical company. The students interested Amgen in testing in microgravity three drugs that were under development—two targeting bone loss and one targeting muscle atrophy. Not only did Amgen provide substantial funding for the three investigations, the company also contributed significant in-kind resources.

Bone remodeling—the natural breakdown and rebuilding of bone—occurs in a balanced fashion in healthy bone so that the rate of rebuilding, known as formation, equals the rate of breakdown and absorption, known as resorption. This cycle of breakdown and buildup helps us to maintain skeletal strength and repair injuries such as fractures, so we can continue to enjoy normal mobility. When this natural process is out of balance, bones and health may suffer.

The first investigation launched in 2001 and looked at using Osteoprotegrin (OPG), while in 2011 researchers flew a sclerostin antibody treatment. OPG and the sclerostin antibody are used as drugs to mitigate bone loss and are based on naturally occurring molecules in the body. The 2007 flight studied myostatin, a preclinical therapy for treating muscle loss. All three therapeutics, which were in preclinical development with Amgen during the time of their flights, had positive impacts on maintaining bone strength.

Moving therapies from the lab to the medicine cabinet takes time, as did the progression of these studies, which spanned a decade in orbit as the space station was under construction. This duration enabled advances in the ways researchers conducted their microgravity investigations, including enhancements to the available tools for analysis.

Now that the station is complete, this research can continue for longer test runs than shuttle missions permitted, thereby increasing data collection and the potential for discovery.

Protein crystals in microgravity

Pharmaceutical companies have developed myriad ways to fight ailments from arthritis to cancer, but the body's immune systems inspired the newest weapons in their arsenal. A new generation of drugs targets specific attackers, sparking immune cells into action. These drugs, called monoclonal antibodies

> The main goal is the development of new, biologically based drugs, providing even tougher weapons for physicians fighting human diseases.

(MABs), are engineered proteins that bind themselves to substances that cause disease. Monoclonal antibodies can be created for almost any target inside a cell or on its surface, allowing greater specificity and fewer side effects than conventional therapies. They include top-selling drugs used to treat several types of inflammation and cancer.

To be effective, though, MABs have to be dispensed in large quantities, which can make them difficult to administer to patients. Rather than simply swallowing a pill, people receiving MAB therapy must receive injections or intravenous infusions. Now research on the International Space Station is changing that. By taking MABs into space and crystallizing them in microgravity, Merck Research Laboratories is working toward high-concentration, high-quality mixtures that can be given to patients more efficiently.

Space is an excellent environment to study complex, three-dimensional proteins, because gravity and convective forces do not get in the way of crystal formation, which allows creation of larger and more perfect crystals. With large crystals, scientists on the

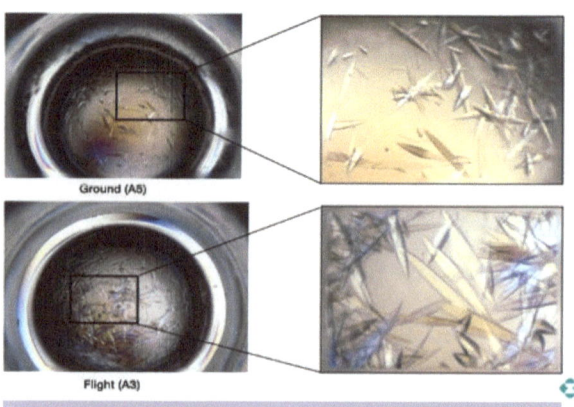

The difference between protein crystals grown on the ground (top) versus in microgravity (bottom).
Image credit: Merck

ground can use X-ray crystallography to determine how the protein is organized. Determining protein structures helps researchers design new drugs.

The first experiment on the space station yielded crystals larger than those that could be grown on the ground, which was promising. Merck continued space-based protein crystallization experiments on 10 separate shuttle flights.

Based on these previous findings, Paul Reichert, chemistry research fellow at Merck, expects to grow uniform suspensions of 5-micron crystals without impurities that can develop on Earth. He also plans to study how temperature gradients can affect the beginning of crystal formation, or nucleation. Reichert flew two different MAB experiments on the SpaceX-3 cargo flight in April 2014 and is planning additional experiments to launch on the SpaceX-6 flight, planned for April 2015.

These high concentrations of protein crystals could improve the way patients receive MABs to treat a wide range of diseases. Because MABs have to be delivered in large amounts, they are usually administered intravenously in a hospital, where a patient would have to wait for several hours to receive the full dose. Highly concentrated suspensions of crystallized proteins, produced in microgravity, could instead be given in a simple shot in a doctor's office. The fluid would look similar to milk, Reichert said, opaque with high concentrations of crystallized MABs making it appear white.

Highly-concentrated MAB mixtures also would be more efficient to ship and store. Currently, different components of MAB treatments are manufactured at different sites and then shipped overseas or to different places in the United States to be formulated into drugs given to patients.

Reichert hopes the work will attract other scientists who want to conduct advanced research in space. But the main goal is the development of new, biologically based drugs, providing even tougher weapons for physicians fighting human diseases.

Muscle atrophy: Mice on the ISS helping life on Earth

Muscles atrophy, or waste away, when they are not used. In microgravity, muscles atrophy even in those who exercise regularly. Without normal gravity to work against, in fact, some muscles begin to atrophy within days after an astronaut reaches orbit. Some people still on Earth experience muscle loss because of diseases or injuries that can make it harder to move, although

> The overall goal is to use ISS to learn more about muscle wasting so scientists and drug companies can help patients who are at risk.

atrophy happens more gradually here. Ultimately, atrophy affects everyone, as all humans lose muscle as they get older.

Scientists still are not sure what happens to trigger muscle loss. Atrophy comes from many pathways in the body, from different cell signaling processes to removal of proteins and loss of amino acids. If scientists can understand how these pathways regulate muscle mass, they might be able to develop new treatments for diseases. First, though, they need a better picture of atrophy itself, and the study of rodents is helping. By observing changes to the genetic activity of mice, scientists at the Novartis Institutes for BioMedical Research (NIBR) hope to learn more about microgravity's effects on muscle mass.

The Novartis team got involved at the invitation of the Center for the Advancement of Science in Space (CASIS), which manages the National Laboratory. Samuel Cadena and his colleagues at Novartis have

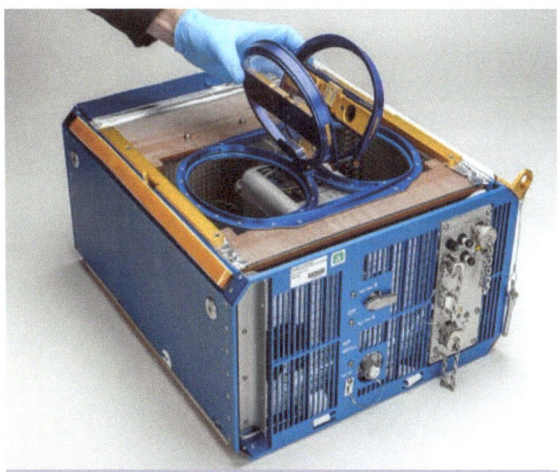

NASA's rodent habitat module seen with both access doors open.
Image credit: NASA

been studying atrophy for several years, and consider using the space station to further their work a unique opportunity.

Cadena and his research partners are studying mice genetically engineered to resist muscle loss. Called knockout mice, these animals lack specific genes that would enable them to make a protein called Muscle Ring Finger-1, or MuRF-1, which hastens muscle loss by labeling certain proteins for degradation. Expression of this protein increases in several muscle atrophy situations, including spaceflight.

One group of 10 MuRF-1 knockout mice flew to the space station on SpaceX-4, arriving on Sept. 23, 2014. A second set stayed at NASA's Kennedy Space Center. Each group was accompanied by an experimental control group of 10 mice whose genes were not changed. The animals stayed in space for three weeks then were euthanized and sent back to Earth for study. Researchers are still analyzing early data, but expect that the knockout mice experienced less atrophy than the control mice, helping to validate the MuRF-1 model.

The overall goal is to learn more about muscle wasting so scientists and drug companies can help patients who are at risk. The partnerships that enable research aboard the orbiting laboratory continue to push the envelope of science in space, seeking answers to propel exploration and benefit people on Earth.

Link to Archived Stories and Videos

http://www.nasa.gov/mission_pages/station/research/benefits/index.html

Authors and Principal Investigators by Section

Principal Investigator (PI) listed for stories focused on a specific space station investigation.

Human Health

Robotic arms lend a healing touch
Author: CSA
PI: Dr. Garnette Sutherland, University of Calgary

Robots from space lead to one-stop breast cancer diagnosis treatment
Author: Jessica Eagan, NASA
PI: Dr. Mehran Anvari, Scientific Director & CEO, Centre for Surgical Invention & Innovation (CSii), Hamilton, Canada.

Improved eye surgery with space hardware
Author: ESA
PI: A. Clarke, Charité Universitätsmedizin, Berlin, Germany

Sensor technologies for high-pressure jobs and operations
Author: ESA
PI: Hans-Christian Gunga, Charité Universitätsmedizin, Berlin, Germany

Bringing space station ultrasound to the ends of the Earth
Author: Mark Wolverton, NASA
PI: Scott Dulchavsky, M.D., Henry Ford Hospital, Detroit, Michigan

Are you asthmatic? Your new helper comes from space
Author: ESA
PI: Lars Karlsson and Lars Gustafsson, Karolinska Institutet, Dept of Physiology and Pharmacology, Stockholm, Sweden

Cold plasmas assist in wound healing
Author: NASA
PI: Dr. Hubertus M. Thomas, Deutsches Zentrum für Luft- und Raumfahrt e.V. (DLR), Germany;
Prof. V.E. Fortov, Institute for High Energy Densities (IHED, RAS), Russia

Preventing bone loss in spaceflight with prophylactic use of bisphosphonate: Health promotion of the elderly by space medicine technologies
Author: Hiroshi Ohshima, JAXA
PI: Adrian Leblanc, United Space Research Association, and Toshio Matsumoto, Tokushima University

Improved scanning technologies and insights into osteoporosis
Author: ESA
PI: Christian Alexandre, M.D., University of St. Etienne, France

Good diet, proper exercise help protect astronauts' bones
Author: Bill Jeffs, NASA
PI: Scott Smith and Jean Sibonga, NASA

Add salt? Astronauts' bones say please don't
Author: ESA
PI: Petra Frings-Meuthen, German Aerospace Center (DLR)

Early detection of immune changes prevents painful shingles in astronauts and in Earth-bound patients
Authors: Satish K. Mehta, Duane L. Pierson, and C. Mark Ott, NASA
PI: Satish K. Mehta, Duane L. Pierson, and C. Mark Ott, NASA

Station immunology insights for Earth and space
Author: Jessica Nimon, NASA
PI: Millie Hughes-Fulford, Director of the Laboratory of Cell Growth at the University of California, San Francisco

Targeted treatments to improve immune response
Author: ESA
PI: M. Maccarrone, N. Battista, University of Teramo, Teramo, Italy

High-quality protein crystal growth experiment aboard Kibo
Authors: Mitsugu Yamada and Kazunori Ohta, JAXA
PI: Over 80 Principal investigators conducted JAXA protein crystal growth experiments on the ISS.

Cancer-targeted treatments from space station discoveries
Author: Laura Niles, NASA
PI: Dennis Morrison, NASA's Johnson Space Center and NuVue Therapeutics, Inc.

Using weightlessness to treat multiple ailments
Authors: I. B. Kozlovskaya and Ye. S. Tomilovskaya, Institute of Biomedical Problems of the Russian Academy of Sciences (IBMP RAS), Moscow, Russia
PI: Ye. S. Tomilovskaya

Microbiology applications from fungal research in space
Author: ESA
PI: D. Hasegan, G. Mogildea, Romanian Institutes of Space Science and Biology, Bucharest, Romania; E. Chatzitheodoridis, National Technical University of Athens, Greece

Plant growth on ISS has global impacts on Earth
Author: Tara Ruttley, NASA
PI: Weijia Zhou, Ph.D., of the Wisconsin Center for Space Automation and Robotics, University of Wisconsin-Madison

Experiments with higher plants on the Russian Segment of the International Space Station
Authors: V. N. Sychev, M. A. Levinskikh, I. G. Podolsky, Institute of Biomedical Problems of the Russian Academy of Sciences (IBMP RAS), Moscow, Russia; G. E. Bingham (Utah State University, Space Dynamics Laboratory, Logan, Utah, USA)
PI: V. N. Sychev, M. A. Levinskikh, I. G. Podolsky, G. E. Bingham
Colleagues from many Russian and non-Russian organizations participated in carrying out work according to the Rasteniya program in the Lada greenhouse on the ISS RS. The contributions of S. A. Gostimsky (M. V. Lomonosov Moscow State University), and M. Sugimoto (Okayama University, Institute of Bioresources, Okayama, Japan) should be especially noted.

Space cardiology for the benefit of health care
Authors: R. M. Baevsky, Ye.Yu. Bersenev, I.I. Funtova, Institute of Biomedical Problems of the Russian Academy of Sciences (IBMP RAS), Moscow, Russia
PI: R. M. Baevsky, Ye.Yu. Bersenev, I.I. Funtova

Biological rhythms in space and on Earth
Authors: NaomuneYamamoto, JAXA and Kuniaki Otsuka, Tokyo Women's Medical University
PI: Chiaki Mukai, JAXA

Innovative space-based device promotes restful sleep on Earth
Authors: R. M. Baevsky, Ye. S. Luchitskaya, I. I. Funtova, Institute of Biomedical Problems of the Russian Academy of Sciences (IBMP RAS), Moscow, Russia
PI: R. M. Baevsky, Ye. S. Luchitskaya, I. I. Funtova

New technology simulates microgravity and improves balance on Earth
Authors: I.B. Kozlovskaya, I.V. Sayenko, Institute of Biomedical Problems of the Russian Academy of Sciences (IBMP RAS), Moscow, Russia
PI: I.B. Kozlovskaya

New ways to assess neurovestibular system health in space also benefits those on Earth
Authors: L. N. Kornilova, I. A. Naumov, G. A. Ekimovskiy, Yu. I. Smirnov, Institute of Biomedical Problems of the Russian Academy of Sciences (IBMP RAS), Moscow, Russia
PI: L. N. Kornilova, I. A. Naumov, Yu. I. Smirnov

Space research leads to non-pharmacological treatment and prevention of vertigo, dizziness and equilibrium disturbances
Authors: L. N. Kornilova, I. A. Naumov, G. A. Ekimovskiy, Institute of Biomedical Problems of the Russian Academy of Sciences (IBMP RAS), Moscow, Russia
PI: L. N. Kornilova, I. A. Naumov, G. A. Ekimovskiy

Capturing the secrets of weightless movements for Earth applications
Authors: Salvatore Pignataro, Gabriele Mascetti, Italian Space Agency
PI: Francesco Lacquaniti, Center of Space Biomedicine, University of Rome Tor Vergata;
Giancarlo Ferrigno, Department of Electronics, Information and Bioengineering

Space technologies in the rehabilitation of movement disorders
Authors: I.V. Sayenko, I.B. Kozlovskaya, Institute of Biomedical Problems of the Russian Academy of Sciences (IBMP RAS), Moscow, Russia
PI: I.V. Sayenko

Earth Observation and Disaster Response

Earth remote sensing from the space station
Author: ISS Program Earth Observations Working Group

Coastal ocean sensing extended mission
Author: Curtiss Davis, College of Earth, Ocean, and Atmospheric Sciences, Oregon State University
PI: Mary E. Kappus, US Naval Research Laboratory; Michael R. Corson, US Naval Research Laboratory

Visual and instrumental scientific observation of the ocean from space
Authors: A. N. Yevguschenko, State Organization "Gagarin Research&Test Cosmonaut Training Center"; B. V. Konovalov, P. P. Shirshov Institute of Oceanography of the Russian Academy of Sciences

Space station camera captures Earthly disaster scenes
Authors: Dauna Coulter and Jaganathan Ranganathan, SERVIR Team, NASA's Marshall Space Flight Center
PI: Burgess Howell, NASA

Clear high-definition images aid disaster response
Authors: Hideaki Shinohara and Sayaka Umemura, JAXA
PI: Chikara Harada, JAXA

Innovative Technology

Advanced ISS technology supports water purification efforts worldwide
Author: Arun Joshi, NASA
PI: Donald L. Carter, Robyn Gatens

Exploring the wonders of fluid motion: Improving life on Earth through understanding the nature of Marangoni convection
Author: Satoshi Matsumoto, JAXA
PI: Hiroshi Kawamura, Tokyo University of Science, Koichi Nishino, Yokohama National University, Shinichi Yoda, JAXA, Yasuhiro Kamotani, Case Western Reserve University, Satoshi Matsumoto, JAXA

Space station-inspired mWater app identifies healthy water sources
Author: Jessica Nimon, NASA
PI: John and Annie Feighery, mWater

Space-tested fluid flow concept advances infectious disease diagnoses
Author: Mike Giannone, NASA Glenn Research Center
PI: Mark Weislogel, NASA

Improving semiconductors with nanofibers
Author: Masato Katsuta, JAXA
PI: Takatoshi Kinoshita, Graduate School of Engineering, Nagoya Institute of Technology

InSPACE's big news in the nano world
Author: Mike Giannone, NASA
PI: Eric M. Furst, University of Delaware

Deploying small satellites from ISS
Author: Hitoshi Morimoto, JAXA
PI: Yusuke Matsumura, JAXA

Pinpointing time and location
Author: ESA
PI: Felix Huber, Steinbeis Transferzentrum Raumfahrt, Gaufelden - Stuttgart, Germany

Space station technology demonstration could boost a new era of satellite-servicing
Author: Adreinne Alessandro, NASA
PI: Frank J. Cepollina, Benjami B. Reed

Cool flame research aboard space station may lead to a cleaner environment on Earth
Author: Mike Giannone, NASA
PI: Folman Williams, Daniel L. Dietrich

Robonaut's potential shines in multiple space, medical and industrial applications
Author: Laura Niles
PI: Myron A. Diftler

Global Education

Inspiring the next generation of students with the International Space Station
Author: Jessica Nimon, NASA
PI: NASA, CSA, ESA, JAXA, Roscosmos

Student scientists receive unexpected results from research in space
Author: Jessica Nimon, NASA
PI: Zahaan Bharmal

Europe's alliance with space droids
Author: ESA
PI: ESA

NASA has HUNCH about student success in engineering
Author: Laura Niles

Tomatosphere™: Sowing the seeds of discovery through student science
Author: CSA
PI: Michael Dixon, University of Guelph

Students photograph Earth from space via Sally Ride EarthKAM program
Author: Arun Joshi, NASA
PI: Sally Ride, Karen Flammer

Try zero G2: Igniting the passion of the next generation in Asia
Author: Yayoi Miyagawa, JAXA
PI: JAXA

Asian students work with astronauts in space missions
Author: Muneo Takaoki, JAXA
PI: Kibo-ABC (Asian Beneficial Collaboration through Kibo Utilization) Initiative

Educational benefits of the space experiment "Shadow-beacon" on ISS
Author: V.A. Strashinskiy, Central Research Institute for Machine Building (FGUP TsNIIMash)
PI: V.A. Strashinskiy, (FGUP TsNIIMash)

Students get fit the astronaut way
Author: Jessica Nimon, NASA
PI: Charles Lloyd, NASA

Inspiring youth with a call to the International Space Station
Authors: Jessica Nimon and Camille Alleyne, NASA
PI: Frank Bauer

Calling cosmonauts from home
Author: S.V.Avdeev, FGUP TsNIIMash
PI: Space experiment "Coulomb Crystal" – V.Ye.Fortov, Joint Institute for High Temperatures of the Russian

Academy of Sciences
Space experiment "Shadow-Beacon" – O.M.Alifanov, Moscow Aviation Institute (National Research University); V.A.Strashinskiy, FGUP TsNIIMash
Space experiment "MAI-75" - O.M.Alifanov, V.K.Odelevskiy; Moscow Aviation Institute (National Research University)
Space experiment "Great Start" – M.Yu.Belyaev, OAO "Rocket and Space Corporation "Energia" after S.P. Korolev"

MAI-75 experiment, main results and prospects for development in education

Authors: O. M. Alifanov, S. O. Firsyuk, V. K. Odelevsky (Moscow Aviation Institute (National Research University)); N. S. Biryukova (Central Research Institute for Machine Building (FGUP TsNIIMash)); S. N. Samburov, A. I. Spirin (OAO Rocket and Space Corporation Energia after S.P. Korolev (RSC-Energia))
PI: O. M. Alifanov, V. K. Odelevsky, Moscow Aviation Institute (National Research University)

Economic Development of Space

Water production in space: Thirsting for a solution
Authors: Kathy Watkins-Richardson, Melissa Gaskill
Company name: UTC Aerospace Systems

Commercialization of low-Earth Orbit (LEO)
Authors: Kathy Watkins-Richardson, Melissa Gaskill
Company name: NanoRacks, LLC

Innovative public-private partnerships for ISS cargo services: Part 1
Authors: Kathy Watkins-Richardson, Melissa Gaskill
Company name: SpaceX

Innovative public-private partnerships for ISS cargo services: Part 2
Authors: Kathy Watkins-Richardson, Melissa Gaskill
Company name: Orbital Sciences Corporation

Precision pointing platform for Earth observations from the ISS
Authors: Kathy Watkins-Richardson, Melissa Gaskill
Company name: Teledyne Brown Engineering, Inc.

The Groundbreaker: Earth observation
Authors: Kathy Watkins-Richardson, Melissa Gaskill
Company name: UrtheCast

A flock of CubeSats photographs our changing planet
Authors: Rebecca Boyle and Melissa Gaskill
Company name: Planet Labs

Stretch your horizons, Stay Curious™
Authors: Kathy Watkins-Richardson, Melissa Gaskill
Company name: Kentucky Space/Space Tango, Inc.

Mission critical: Flatworm experiment races the clock after splashdown
Author: Craig Simon, President and CEO
Company name: FedEx SupplyChain

Economic development of space in JAXA
Authors: Sayaka Umemura and Kazuyuki Tasaki, JAXA
Company and organization names: Mitsubishi Electric (MELCO)/Orbital Sciences Corporation, Japan Space Forum, Hamamatsu Photonics K.K., Fuji Chemical Co., Ltd., Shiseido Co., Ltd., Yakult Honsha Co., Ltd., Japan Manned Space Systems Corporation (JAMSS)/Technological Institute of Aeronautics (ITA)/Brazilian Space Agency (AEB)/Brazilian National Institute for Space Research (INPE), Interprotein Corporation, Chugai Pharmaceutical Co., Ltd., ARKRAY, and Junkosha Inc.

Colloids in space: Where consumer products and science intersect
Authors: Kathy Watkins-Richardson, Melissa Gaskill
PI: David A. Weitz, Harvard University; Matthew Lynch, Proctor and Gamble

Space mice teach us about muscle and bone loss
Authors: Rebecca Boyle, Kathy Watkins-Richardson and Melissa Gaskill
PI: H. Q. Han, Amgen Research

Protein crystals in microgravity
Authors: Rebecca Boyle and Melissa Gaskill
PI: Paul Reichert, Merck Research Laboratories

Muscle atrophy: Mice on the ISS helping life on Earth
Authors: Rebecca Boyle, Kathy Watkins-Richardson and Melissa Gaskill
PI: Samuel Cadena, Novartis Institutes for Biomedical Research

www.ingramcontent.com/pod-product-compliance
Lightning Source LLC
Chambersburg PA
CBHW041313180526
45172CB00004B/1077